The Geometry of Special Relativity

Textbooks in Mathematics
Series editors:
Al Boggess, Kenneth H. Rosen

https://www.routledge.com/Textbooks-in-Mathematics/book-series/CANDHTEXBOOMTH

The Geometry of
Special Relativity
Second Edition

Tevian Dray

Department of Mathematics, Oregon State University

CRC Press

Taylor & Francis Group

Boca Raton London New York

CRC Press is an imprint of the
Taylor & Francis Group, an **informa** business

A CHAPMAN & HALL BOOK

The figures in Chapters 1-13 and 16-17 originally appeared in the online text [Dray 10] and are reproduced with permission under the Creative Commons by-nc-nd license.

Second edition published 2021
by CRC Press
6000 Broken Sound Parkway NW, Suite 300, Boca Raton, FL 33487-2742

and by CRC Press
2 Park Square, Milton Park, Abingdon, Oxon, OX14 4RN

© 2021 Taylor & Francis Group, LLC

First edition published by CRC Press 2012

CRC Press is an imprint of Taylor & Francis Group, LLC

Library of Congress Cataloging-in-Publication Data

Names: Dray, Tevian, author.
Title: The geometry of special relativity / Tevian Dray, Department of
Mathematics, Oregon State University.
Description: Second edition. | Boca Raton : Chapman & Hall/CRC Press, 2021.
| Series: Advances in applied mathematics | Includes bibliographical
references and index.
Identifiers: LCCN 2021000964 (print) | LCCN 2021000965 (ebook) | ISBN
9781032008202 (hardback) | ISBN 9781138063921 (paperback) | ISBN
9781351663212 (ebook)
Subjects: LCSH: Special relativity (Physics) | Space and time--Mathematical
models.
Classification: LCC QC173.65 .D73 2021 (print) | LCC QC173.65 (ebook) |
DDC 530.11--dc23
LC record available at https://lccn.loc.gov/2021000964
LC ebook record available at https://lccn.loc.gov/2021000965

ISBN: 9781032008202 (hbk)
ISBN: 9781138063921 (pbk)
ISBN: 9781351663212 (ebk)

Lorentz transformations are just hyperbolic rotations.

Contents

Preface

Preface to the First Edition

The unification of space and time introduced by Einstein's special theory of relativity is one of the cornerstones of the modern scientific description of the universe. Yet the unification is counterintuitive because we perceive time very differently from space. Even in relativity, time is not just another dimension, it is one with different properties. Some authors have tried to "unify" the treatment of time and space, typically by replacing t by it, thus hiding some annoying minus signs. But these signs carry important information: Our universe, as described by relativity, is *not* Euclidean.

This short book treats the geometry of hyperbolas as the key to understanding special relativity. This approach can be summarized succinctly as the replacement of the ubiquitous γ symbol of most standard treatments with the appropriate hyperbolic trigonometric functions. In most cases, this not only simplifies the appearance of the formulas, but emphasizes their geometric content in such a way as to make them almost obvious. Furthermore, many important relations, including but not limited to the famous relativistic addition formula for velocities, follow directly from the appropriate trigonometric addition formulas.

I am unaware of any other introductory book on special relativity that adopts this approach as fundamental. Many books point out the relationship between Lorentz transformations and hyperbolic rotations, but few actually make use of it. A pleasant exception is the original edition of Taylor and Wheeler's marvelous book [Taylor and Wheeler 63], but much of this material is not included in the second edition [Taylor and Wheeler 92].

This book is not intended as a replacement for any of the excellent textbooks on special relativity. Rather, it is intended as an introduction to a particularly beautiful way of looking at special relativity, in hopes of encouraging students to see beyond the formulas to the deeper structure. Enough applications are included to get the basic idea, but these would probably need to be supplemented for a full course.

Much of the material presented can be understood by those familiar with the ordinary trigonometric functions, but occasional use is made of elementary differential calculus. In addition, the chapter on electricity and magnetism assumes the reader has seen Maxwell's equations and has at least a passing acquaintance with vector calculus. A prior course in calculus-based physics, up to and including electricity and magnetism, should provide the necessary background.

Preface to the Second Edition

The core of the book remains unchanged from the first edition, with its short, descriptive, accessible treatment of the material. However, in response to reader feedback, the initial treatment of Minkowski space and spacetime diagrams has been expanded slightly, with minor changes in notation. In addition, several new topics have been added, numerous small typos have been corrected, and more homework problems have been added.

The primary notational change made was to eliminate the use of β for hyperbolic angles, in favor of α, since other authors often use β as shorthand for $\frac{v}{c}$. Where already present, α was changed to ψ.

The new material includes a geometric derivation of Lorentz transformations in Section 5.5, a discussion of three-dimensional spacetime diagrams in Chapter 14, including several examples related to homework problems, and a brief geometric description of "area" in Chapter 15, and how it can be used to measure time and distance.

Overview

After a general introduction in Chapter 1, the basic physics of special relativity is described in Chapter 2. This is a quick, intuitive introduction to special relativity, and it sets the stage for the geometric treatment that follows. Chapter 3 summarizes some standard (and some not so standard) properties of ordinary two-dimensional Euclidean space, expressed in terms of the usual circular trigonometric functions; this geometry is referred to as *circle geometry*. This material has deliberately been arranged to closely parallel the treatment of two-dimensional Minkowski space in Chapter 4 in terms of hyperbolic trigonometric functions, which we call *hyperbola geometry*.[1] Special relativity is covered again from the geometric point of view in Chapters 5 and 6, followed by a discussion of some of the standard paradoxes in Chapter 8, applications to relativistic mechanics in Chapter 9, and the relativistic unification of electricity and magnetism in Chapter 11. Chapters 7, 10, and 12 present relevant problems and their solutions. Chapter 13 concludes our tour of relativity with a brief discussion of the further steps leading to Einstein's theory of general relativity. As noted above, two new chapters discuss three-dimensional spacetime diagrams (Chapter 14) and the use of double-null "light boxes" to measure spacetime intervals (Chapter 15). The final two chapters discuss applications of hyperbola geometry in mathematics, treating hyperbolic geometry in Chapter 16, and giving a geometric construction of the derivatives of trigonometric functions, as well as the exponential function and its derivative, in Chapter 17.

---

[1] Hyperbola geometry should not be confused with *hyperbolic geometry*, the curved geometry of the two-dimensional unit hyperboloid; see Chapter 16.

Acknowledgments

This book started out as class notes [Dray 03] for a course on *Reference Frames*, which in turn forms part of a major upper-division curriculum reform effort titled *Paradigms in Physics*, which was begun in the Department of Physics at Oregon State University in 1997. The class notes were subsequently published online in Wiki format [Dray 10], and were used for 15 years as the primary text in that course; an abbreviated version was also published as a journal article [Dray 04].

I am grateful to all of the faculty involved in this effort, but especially to the leader of the project, my wife, Corinne Manogue, for support and encouragement at every stage. The *Paradigms in Physics* project has been supported in part by NSF grants DUE–965320, 0231194, 0618877, 1023120, 1323800, 1836603, and 1836604, supplemented with funds from Oregon State University; my own participation was made possible thanks to the (sometimes reluctant) support of my department chair at the time, John Lee. I was fortunate in having excellent teaching assistants, Jason Janesky and Emily Townsend, when I first taught the course.

A course based on an early draft of this book was taught at Mount Holyoke College in 2002, giving me an opportunity to make further revisions; my stay at Mount Holyoke was partially supported by their Hutchcroft Fund. I am grateful to Greg Quenell for having carefully read the manuscript at that time and suggesting improvements, and to Alex Brummer, who did the same ten years later. More recently, this material has been taught at Oregon State University as part of a newly designed, sophomore-level course in mechanics by Liz Gire, who also provided extensive, constructive feedback.

I thank David Griffiths, an old friend (and Corinne's former teacher), for permission to "borrow" homework problems from his standard textbook [Griffiths 99], as well as to include solutions using hyperbola geometry. Thanks also to Rob Salgado for permission to discuss his geometric approach to measurement using causal diamonds [Salgado 16].

A special thank you is due to the readers of the first edition who pointed out typos and made suggestions for improvement.

Last, but not least, I thank the many students who struggled to learn physics from a mathematician, enriching all of us.

List of Figures

List of Tables

1

Introduction

1.1 Newton's Relativity

Our daily experience leads us to believe in Newton's laws. When you drop a ball, it falls straight down. When you throw a ball, it travels in a uniform (compass) direction—and falls down. We appear to be in a constant gravitational field, but apart from that no forces act on the ball. This isn't the full story, of course, because we are ignoring such things as air resistance and the spin of the ball. Nevertheless, Newton's laws seem to give a pretty good description of what we observe, and so we base our intuitive understanding of physics on them.

But Newton's laws are wrong.

Yes, gravity is more complicated than this simple picture. The earth's gravitational field isn't really constant. Furthermore, other nearby objects act gravitationally on us, notably the moon. As discussed in the final chapter of this book, this action causes tides.

A bigger problem is that the earth is round. Anyone who flies from San Francisco to New York is aware that due east is *not* a straight line, defined in this case as the shortest distance between two points. In fact, if you travel in a straight line that (initially) points due east from my home in Oregon, you will eventually pass to the south of the southern tip of Africa![1]

So east is not east.

But the real problem is that the earth is rotating. Try playing catch on a merry-go-round! Balls certainly don't seem to travel

---

[1] You can check this by stretching a string on a globe so that it goes all the way around, is as tight as you can make it, and goes through Oregon in an east/west direction.

in a straight line! Newton's laws don't work here, and, strictly speaking, they don't work on (that is, in the reference frame of) the earth's surface. The motion of a Foucault pendulum can be thought of as a Coriolis effect caused by an external *pseudoforce*. Also, a plumb bob doesn't actually point toward the center of the earth!

So down is not down.

1.2 Einstein's Relativity

All of the problems above come from the fact that, even without worrying about gravity, the earth's surface is *not* an inertial frame. An inertial frame is, roughly speaking, one in which Newton's laws do hold. Playing catch on a train is little different from on the ground—at least in principle, as long as the train is not speeding up or slowing down. Furthermore, an observer on the ground would see nothing out of the ordinary, it merely being necessary to combine the train's velocity with that of the ball.

However, shining a flashlight on a moving train—especially the description of this from the ground—turns out to be another story, which we study in more detail in Chapters 2 and 6. Light doesn't behave the way balls do, and this difference forces a profound change in our description of the world around us. As we will see, this forces moving objects to change in unexpected ways: Their clocks slow down, they change size, and, in a certain sense, they get heavier.

So time is not time.

Of course, these effects are not very noticeable in our daily lives, any more than Coriolis forces affect a game of catch. But some modern conveniences, notably global positioning technology, are affected by relativistic corrections.

The bottom line is that the reality is quite different from what our intuition says it ought to be. The world is neither Euclidean nor Newtonian. Special relativity isn't just some bizarre theory; it is a correct description of nature (ignoring gravity). It is also a beautiful theory, as I hope you will agree. Let's begin.

2

The Physics of
Special Relativity

In which it is shown that time is not the same for all observers.

2.1 Observers and Measurement

This chapter provides a very quick introduction to the physics of
special relativity, intended as a review for those who have seen
it before, and as an overview for those who have not. It is not a
prerequisite for the presentations in subsequent chapters.

Special relativity involves comparing what different observers
see. But we need to be careful about what these words mean.

A *reference frame* is a way of labeling each event with its location in space and the time at which it occurs. Making a measurement corresponds to recording these labels for a particular event.
When we say that an *observer* "sees" something, what we really
mean is that a particular event is recorded in the reference frame
associated with the observer. This has nothing to do with actually
seeing anything—a much more complicated process that would in-
volve keeping track of the light reflected into the observer's eyes.
Rather, an "observer" is really an entire army of observers who
record any interesting events and an "observation" consists of re-
constructing from their journals what took place.

2.2 The Postulates of Special Relativity

Postulate I is the most fundamental postulate of relativity.

Postulate I: The laws of physics apply in all inertial reference frames.

The first ingredient here is a class of preferred reference frames. Simply put, an inertial reference frame is one without external forces. More precisely, an inertial frame is one in which an object initially at rest will remain at rest. Because of gravity, inertial frames must be in free fall; consider a spaceship with its drive turned off or a falling elevator. Gravity causes additional complications, such as tidal effects, which force such freely falling frames to be small (compared to, say, the earth); we revisit this concept in Chapter 13. But special relativity describes a world without gravity, so in practice we describe inertial frames in terms of relative motion at constant velocity, typically in the form of an idealized train.

Applied to mechanics, Postulate I is the principle of *Galilean relativity*. For instance, consider a ball thrown to the right with speed u. Ignoring forces such as gravity and air friction, the ball keeps moving at the same speed forever, since there are no forces acting on it. Try the same thing on a train, which is itself moving to the right with speed v. Then Galilean relativity leads to the same conclusion: As seen from the train, the ball moves to the right with speed u forever. An observer on the ground, of course, sees the ball move with speed $u + v$; Galilean relativity insists only that both observers observe the same physics, namely, the lack of acceleration due to the absence of any forces, but not necessarily the same speed. This situation is shown in Figure 2.1.

Einstein generalized Postulate I by applying it not just to mechanics but also to electrodynamics. However, Maxwell's equations, presented in roughly their present form in the early 1860s, make explicit reference to the speed of light. In MKS units, Gauss's law (see Equation (11.72)) involves the permittivity constant ϵ_0, and Ampère's Law (see Equation (11.75)) involves the permeability constant μ_0; both of these can be measured experimentally.

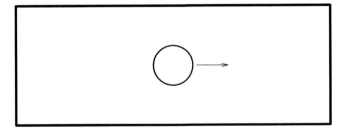

FIGURE 2.1. A passenger on a train throws a ball to the right. On an ideal train, it makes no difference whether the train is moving.

But Maxwell's equations predict electromagnetic waves—including light—with a speed (in vacuum) of

$$c = \frac{1}{\sqrt{\epsilon_0 \mu_0}}. \tag{2.1}$$

Thus, from some relatively simple experimental data, Maxwell's equations *predict* that the speed of light in vacuum is

$$c = 3 \times 10^8 \text{ m/s}. \tag{2.2}$$

The famous Michelson-Morley experiment in 1887, at what is now Case Western Reserve University, set out to show that this speed is relative to the *ether*, so we should be able to measure our own motion relative to the ether by measuring direction-dependent variations in c. Instead, the experiment showed that there were no such variations; Einstein argued that there is therefore no ether![1] Postulate I together with Maxwell's equations therefore lead to

Postulate II: The speed of light is the same for all inertial observers.

Postulate II says not only that the speed of light is constant, but also that this constant must have the same value for all inertial observers. The theory of special relativity follows from Postulates I

---

[1]Of historical interest is the fact that this was not the interpretation given by Michelson and Morley, who instead argued in favor of a now-discredited theory involving the dragging of the ether by the earth. Nonetheless, in 1907 Albert Abraham Michelson became the first American to win the Nobel Prize in Physics, in part due to this work.

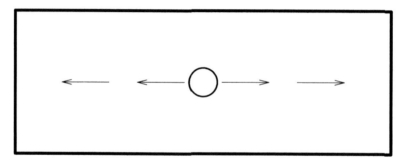

FIGURE 2.2. The same situation as for Figure 2.1, with the ball replaced by a lamp in the exact middle of the train. The light from the
lamp reaches both ends of the train at the same time, whether the train
is moving or at rest.

and II, which were introduced by Einstein in 1905 [Einstein 1905].[2]
As we show next, an immediate consequence of these postulates is
that two inertial observers disagree about whether two events are
simultaneous.

Thought experiments have been around since at least the time
of Euclid. However, the German phrase *Gedankenexperiment* was
not coined until the early 1800s, and was first translated into English in the late 1800s, not long before Einstein's discovery of special relativity. Einstein himself attributed his breakthrough to
thought experiments, including a childhood attempt to imagine
what one would see if one could ride along with a beam of light.
However, it remains unclear exactly what path he took to reach his
conclusions about the observer-dependence of simultaneity. One
possible path [Norton 06] is the following thought experiment.

Consider a train at rest with a lamp in the middle, as shown
in Figure 2.2. After the light is turned on, light reaches both ends
of the train at the same time, having traveled in both directions
at constant speed *c*. Now try the same experiment on a moving
train. This is still an inertial frame, and so, just as with the ball
in the previous example, one obtains the same result, namely, that
the light reaches both ends of the train at the same time *as seen*

---

[2]Albert Einstein won the Nobel Prize in Physics in 1921, but not for his
work on special relativity.

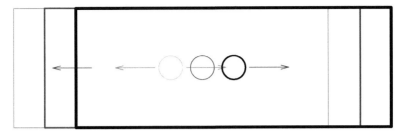

FIGURE 2.3. An observer on the ground sees the train go past with its lamp, as in Figure 2.2. Here, and in subsequent figures, different shades of gray represent the train at different times. Since the ground-based observer must also see the light travel with speed c, and since the ends of the train are moving while the light is traveling, this observer concludes that the light reaches the rear of the train before it reaches the front.

by an observer on the train. However, the second postulate leads to a very different result for an observer on the ground. According to this postulate, the light travels at speed c as seen from the ground, *not* the expected $c \pm v$. But, as seen from the ground, the ends of the train also move while the light is getting from the middle of the train to the ends. The rear wall "catches up" with the approaching light beam, whereas the front wall "runs away." As shown in Figure 2.3, the net result is that the ground-based observer sees the light reach the rear of the train before it reaches the front; these two observers therefore disagree about whether the light reaches both ends of the train simultaneously.

2.3 Time Dilation and Length Contraction

We have seen that the postulates of relativity force the surprising conclusion that time is observer-dependent. We now examine this phenomenon in more detail.

Consider again a train, of height h, with a beam of light bouncing up and down between mirrors on the floor and ceiling, as shown in Figure 2.4(a). The time between bounces can be interpreted as the "ticks" of a clock, and this interval, *as measured on the train*, is independent of whether the train is moving. However, as shown in Figure 2.4(b), a stationary observer on the ground sees something quite different.

From the ground, the light appears to move diagonally, and hence travels a longer path than the vertical path seen on the train. But since the light must move at the same speed for both observers, each "tick" takes longer according to the ground-based observer than it does for the observer on the train. Thus, the observer at rest sees the "clock" of the moving observer run slow.

If we work through each step of this argument carefully, the key assumption is Postulate II—namely, that the light must travel at the same speed for both observers. But this is not the behavior we expect from our daily experience.

To compute how the times are related, we must first introduce some notation. Let t denote time as measured on the ground, and t' denote time as measured on the train; we similarly use x and x' to measure length. One tick of the clock as seen on the train takes time $\Delta t'$, where the distance traveled is

$$h = c\,\Delta t'. \tag{2.3}$$

Suppose the same tick takes time Δt as seen from the ground. In this time, the light travels a distance $c\,\Delta t$, which is the hypotenuse of a right triangle with legs h and $v\,\Delta t$. As shown in Figure 2.5, the Pythagorean theorem now leads to

$$(c\,\Delta t)^2 = (v\,\Delta t)^2 + h^2, \tag{2.4}$$

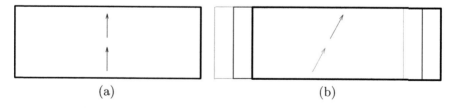

(a)   (b)

FIGURE 2.4. A beam of light bounces up and down between mirrors on the floor and ceiling of a moving train. The time between bounces can be used as a unit of time, but what is seen by (a) a moving observer on the train and (b) a stationary observer on the ground is quite different.

FIGURE 2.5. The (ordinary) Pythagorean theorem for a bouncing light beam on a moving train.

which can be solved for h. Comparing this result with that of Equation (2.3), we obtain

$$\Delta t = \frac{1}{\sqrt{1 - \frac{v^2}{c^2}}} \Delta t', \qquad (2.5)$$

which indeed shows that the moving clock runs slower than the stationary one ($\Delta t > \Delta t'$), at least for speeds $v < c$. The factor relating these times shows up so often that we give it a special symbol,

$$\gamma = \frac{1}{\sqrt{1 - \frac{v^2}{c^2}}}. \qquad (2.6)$$

This effect, called *time dilation*, has important consequences for objects traveling at a significant fraction of the speed of light, but has virtually no effect for objects at everyday speeds ($v \ll c$). This is the effect that allows cosmic rays to reach the earth—the particles' lifetimes, as measured by their own clocks, is many orders of magnitude shorter than the time we observe them traveling through the atmosphere.

Time is not the only thing on which observers do not agree. An observer on the train can measure the length L' of the train station it passes through by multiplying the time $\Delta t'$ it takes the train to pass through the station by its speed v, so that

$$L' = v\,\Delta t'. \tag{2.7}$$

The analogous observation on the ground requires two observers, one at each end of the station, who record the times t_1 and t_2 when the observer reaches their location. They therefore determine the length L of the station to be

$$L = v\,\Delta t \tag{2.8}$$

where $\Delta t = t_2 - t_1$. Since we have just shown that the clocks of moving observers run slow ($\Delta t > \Delta t'$), these lengths must also differ ($L > L'$). More precisely, combining (2.5), (2.7), and (2.8), we obtain

$$L = \gamma L'. \tag{2.9}$$

Thus, the station appears shorter to the moving observer than it really is.

This result can also be obtained by considering a beam of light bouncing horizontally between the front and back of the train, as shown in Figure 2.6(a). As seen on the train, if it takes time $\Delta t'$ to make a round trip, and the length of the train is $\Delta x'$, then we must have

$$c\,\Delta t' = 2\,\Delta x'. \tag{2.10}$$

What does the observer on the ground see? Don't forget that the train is moving, so that, as shown in Figure 2.6(b), the distance traveled in one direction is different from that in the other. More precisely, light starting from the back of the train must "chase" the front; if it takes time Δt_1 to catch up, then the distance traveled is $\Delta x + v\,\Delta t_1$, the sum of the length of the train (as seen from the ground) and the distance the front of the train traveled while the light was under way. Similarly, if the time taken on the return journey is Δt_2, then the distance traveled is $\Delta x - v\,\Delta t_2$. Thus,

$$c\,\Delta t_1 = \Delta x + v\,\Delta t_1 \tag{2.11}$$

and

$$c\,\Delta t_2 = \Delta x - v\,\Delta t_2, \tag{2.12}$$

(a)

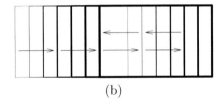
(b)

FIGURE 2.6. A beam of light bounces back and forth between mirrors at the front and back of a moving train. The time of the roundtrip journey can be used to measure the length of the train, but a moving observer on the train (a) and a stationary observer on the ground (b) obtain different results. (As before, different shades of gray represent the moving train at different times, each corresponding to one of the arrows, showing the beam of light at that time.)

or equivalently,

$$\Delta t_1 = \frac{\Delta x}{c - v} \tag{2.13}$$

and

$$\Delta t_2 = \frac{\Delta x}{c + v}. \tag{2.14}$$

Combining these results leads to

$$c\,\Delta t = c\,\Delta t_1 + c\,\Delta t_2$$
$$= \frac{2\,\Delta x}{1 - \frac{v^2}{c^2}}. \tag{2.15}$$

Combining Equations (2.15), (2.5), and (2.10) leads to

$$\Delta x = \sqrt{1 - \frac{v^2}{c^2}}\,\Delta x' = \frac{1}{\gamma}\,\Delta x'. \tag{2.16}$$

Thus, in agreement with (2.9), a moving object appears to be shorter in the direction of motion than it would be at rest. And it doesn't matter whether the object or the observer is moving; it is the relative motion that matters. This effect is known as *length contraction*.

2.4 Lorentz Transformations

Suppose observer O' is moving to the right with speed v. If x denotes the distance of an object from a stationary object O,[3] then the distance between the object and O' *as measured by O* will be $x - vt$. This distance *to* a moving observer behaves like the length *of* a moving object, so we can apply the formula for length contraction (2.16) in the form $\Delta x' = \gamma \, \Delta x$, resulting in

$$x' = \gamma \, (x - vt). \tag{2.17}$$

By Postulate I, the framework used by each observer to describe the other must be the same. In particular, if we interchange the roles of O and O', nothing else should change—except for the fact that the relative velocity (of O with respect to O') is now $-v$ instead of v. By symmetry, we therefore have immediately that

$$x = \gamma \, (x' + vt'), \tag{2.18}$$

where we have been careful not to assume that $t' = t$. In fact, combining Equations (2.17) and (2.18) quickly yields

$$t' = \gamma \left(t - \frac{v}{c^2} x \right) \tag{2.19}$$

and a similar expression for t in terms of x' and t'.

Equations (2.17) and (2.19) together give the *Lorentz transformation* from the frame of the observer at rest (O) to the frame of the moving observer (O').[4] By symmetry, we can invert these expressions simply by replacing v with $-v$.

2.5 Addition of Velocities

Suppose that, as seen from O, O' is moving to the right with speed v and that an object is moving to the right with speed u.

---

[3]This distance need not be constant, but could be a function of time.

[4]Lorentz transformations are named after Hendrik Lorentz, a Dutch physicist who won the 1902 Nobel Prize in Physics.

According to Galileo, we would simply add velocities to determine the velocity of the object as seen from O':

$$u = u' + v. \qquad (2.20)$$

Equation (2.20) can be derived by differentiating the Galilean transformation

$$x = x' + vt, \qquad (2.21)$$

thus obtaining

$$\frac{dx}{dt} = \frac{dx'}{dt} + v. \qquad (2.22)$$

To derive the relativistic formula for the addition of velocities, we proceed similarly. However, since *both* x and t transform, it is useful to use the differential form of the Lorentz transformations, namely,

$$dx = d\left(\gamma\left(x' + vt'\right)\right) = \gamma\left(dx' + v\,dt'\right), \qquad (2.23)$$

$$dt = d\left(\gamma\left(t' + \frac{v}{c^2}x'\right)\right) = \gamma\left(dt' + \frac{v}{c^2}dx'\right). \qquad (2.24)$$

Dividing expression (2.23) by (2.24) leads to

$$\frac{dx}{dt} = \frac{\frac{dx'}{dt'} + v}{1 + \frac{v}{c^2}\frac{dx'}{dt'}}, \qquad (2.25)$$

or, equivalently,

$$u = \frac{u' + v}{1 + \frac{u'v}{c^2}}. \qquad (2.26)$$

Equation (2.26) is known as the *Einstein addition formula*, and shows that the addition of relativistic velocities does not obey our intuitive notion of how velocities should add. It is worth considering special cases of this formula, such as when both u' and v are small compared to c, or when one of them is equal to c. We provide a geometric explanation for these apparently peculiar properties in Chapter 6.

2.6   The Interval

Direct computation using the Lorentz transformations shows that

$$x'^2 - c^2 t'^2 = \gamma^2 \left(x - vt\right)^2 - \gamma^2 \left(ct - \frac{v}{c}x\right)^2$$
$$= x^2 - c^2 t^2, \qquad\qquad (2.27)$$

so that the quantity $x^2 - c^2 t^2$, known as the *interval*, does not depend on the observer who computes it. We explore this concept further in later chapters.

This concludes our brief introduction to the underlying physics. In subsequent chapters we will recover these results from a more geometric point of view.

3

Circle Geometry

In which some standard properties of two-dimensional Euclidean geometry are reviewed, and some more subtle properties are pointed out.

3.1 The Geometry of Trigonometry

We begin with a review of the geometry of trigonometry. Before reading further, you are encouraged to reflect briefly on the most important features of trigonometry. Make a list! As you then read through this chapter, compare your list to the properties discussed here. At the end of the chapter, ask yourself whether your list has changed. (You may wish to repeat this exercise as you read Chapter 4.)

3.2 Distance

The key concept in Euclidean geometry is the *distance function,* which measures the distance between two points. In two dimensions, the squared distance between the point B with coordinates (x, y) and the origin is given by

$$r^2 = x^2 + y^2, \tag{3.1}$$

which is, of course, just the Pythagorean theorem; see Figure 3.1.

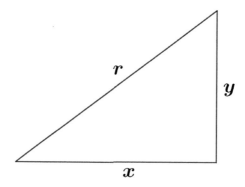

FIGURE 3.1. Measuring distance in Euclidean geometry by using the Pythagorean theorem, which states that $r^2 = x^2 + y^2$.

3.3   Circle Trigonometry

It is natural to study the set of points that are a constant distance from a given point, which of course form a *circle*.

Consider a point P on the circle of radius r shown in Figure 3.2. The angle θ between the line from the origin to P and the (positive) x-axis is *defined* as the ratio of the length the arc of a circle between P and the point $(r, 0)$ to the radius r.[1] Denoting the coordinates of P by (x, y), the basic (circular) trig functions are then *defined* by

$$\cos \theta = \frac{x}{r}, \tag{3.2}$$

$$\sin \theta = \frac{y}{r}, \tag{3.3}$$

$$\tan \theta = \frac{\sin \theta}{\cos \theta}, \tag{3.4}$$

and the fundamental identity

$$\cos^2 \theta + \sin^2 \theta = 1 \tag{3.5}$$

---

[1]This definition is commonly stated in terms of the *unit* circle, a practice we avoid in order to more easily distinguish dimensionless quantities such as angles from dimensionful quantities such as lengths.

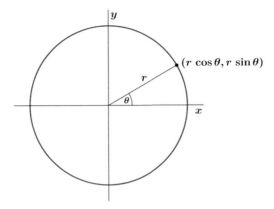

FIGURE 3.2. Defining the (circular) trigonometric functions via the unit circle.

follows from the definition of a circle. We then have the well-known addition formulas[2]

$$\sin(\theta + \phi) = \sin\theta\cos\phi + \cos\theta\sin\phi, \tag{3.6}$$

$$\cos(\theta + \phi) = \cos\theta\cos\phi - \sin\theta\sin\phi, \tag{3.7}$$

$$\tan(\theta + \phi) = \frac{\tan\theta + \tan\phi}{1 - \tan\theta\tan\phi}, \tag{3.8}$$

as well as the derivative formulas[3]

$$\frac{d}{d\theta}\sin\theta = \cos\theta, \tag{3.9}$$

$$\frac{d}{d\theta}\cos\theta = -\sin\theta. \tag{3.10}$$

A unified summary of the (circular) trigonometric functions is shown in Figure 3.3.

---

[2]A particularly elegant proof of Equation (3.7) is to compute the dot product of two unit vectors both algebraically and geometrically, and compare the results. Equations (3.6) and (3.8) follow easily from (3.7); see, for example, [Dray and Manogue 06].

[3]An alternate derivation of these derivative formulas is given in Section 17.1.

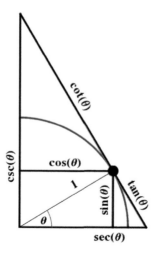

FIGURE 3.3. A unified summary of the (circular) trigonometric functions.

3.4 Triangle Trigonometry

An important class of trigonometric problems involves determining, say, $\cos\theta$ if $\tan\theta$ is known. One can, of course, do this algebraically by using the identity

$$\cos^2\theta = \frac{1}{1 + \tan^2\theta}. \tag{3.11}$$

It is often easier to do this geometrically, as illustrated by the following example.

Suppose you know $\tan\theta = \frac{3}{4}$, and you wish to determine $\cos\theta$. Draw *any* triangle containing an angle whose tangent is $\frac{3}{4}$. In this case, the obvious choice would be the triangle shown in Figure 3.4, with sides of 3 and 4. What then is $\cos\theta$? The hypotenuse clearly has length 5, so $\cos\theta = \frac{4}{5}$.

Trigonometry is not merely about ratios of sides, it is also about projections. Another common use of triangle trigonometry is to determine the sides of a triangle given the hypotenuse r and one angle θ. The answer, of course, is that the sides are $r\cos\theta$ and $r\sin\theta$, as shown in Figure 3.5.

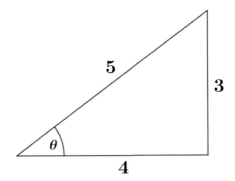

FIGURE 3.4. A right triangle with $\tan\theta = \frac{3}{4}$.

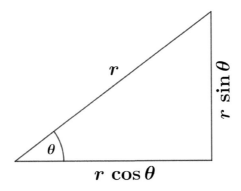

FIGURE 3.5. A right triangle in which the hypotenuse and one angle are known.

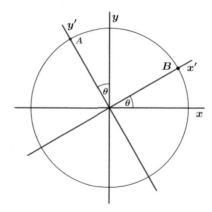

FIGURE 3.6. A rotated coordinate system.

3.5   Rotations

Now consider a new set of coordinates, call them (x', y'), based on axes rotated counterclockwise through an angle θ from the original ones, as shown in Figure 3.6. For instance, the y-axis could point toward true north, whereas the y'-axis might point toward magnetic north. In the primed coordinate system, point B has coordinates $(r, 0)$, and point A has coordinates $(0, r)$, whereas the unprimed coordinates of B and A are, respectively, $(r \cos \theta, r \sin \theta)$ and $(-r \sin \theta, r \cos \theta)$. A little work shows that the coordinate systems are related by a rotation matrix

$$\begin{pmatrix} x \\ y \end{pmatrix} = \begin{pmatrix} \cos \theta & -\sin \theta \\ \sin \theta & \cos \theta \end{pmatrix} \begin{pmatrix} x' \\ y' \end{pmatrix} \qquad (3.12)$$

(where we are assuming that all coordinates are measured in the same units). Furthermore, both coordinate systems lead to the same result for the squared distance, since

$$x^2 + y^2 = x'^2 + y'^2. \qquad (3.13)$$

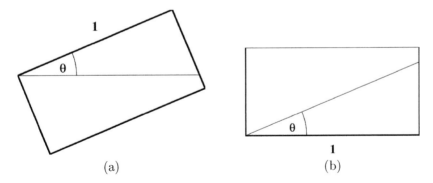

FIGURE 3.7. Width is coordinate-dependent. The rectangle is oriented (a) parallel to the primed axes of Figure 3.6, and (b) parallel to the unprimed axes.

3.6   Projections

Consider the rectangular object of width 1 meter shown in Figure 3.7(a), which has been rotated so it is parallel to the *primed* axes in Figure 3.6. How wide is it? As worded, this question is poorly posed. If width means "extent in the x' direction," then the answer is 1 meter. If, however, width means "extent in the x direction," then the answer is obtained by measuring the *horizontal* distance between the sides of the rectangle, which results in a value *larger* than 1. (The exact value is easily seen to be $1/\cos\theta$.)

Repeat this exercise in the opposite direction. Take the same rectangular object, but orient it parallel to the *unprimed* axes, as shown in Figure 3.7(b). How wide is it? Clearly the "unprimed" width is 1 meter, and the "primed" width is larger (again given by $1/\cos\theta$).

In one of the cases above, the "primed" width is smaller, yet in the other the "unprimed" width is smaller. What is happening here? If you turn your suitcase at an angle, it is *harder* to fit under your seat! It has, in effect, become "longer." But which orientation is best depends, of course, on the orientation of the seat.

Remember this discussion when we address the corresponding questions in relativity in subsequent chapters.

3.7   Addition Formulas

Consider the line through the origin that makes an angle ϕ with the positive x-axis, as shown in Figure 3.8(a). What is its slope? The equation of the line is

$$y = x \tan \phi, \tag{3.14}$$

so that the slope is $\tan \phi$, at least in "unprimed" coordinates. Consider now the line through the origin shown in Figure 3.8(b), which makes an angle ϕ with the positive x'-axis. What is its slope? In "primed" coordinates, the equation of the line is just

$$y' = x' \tan \phi, \tag{3.15}$$

so that, in these coordinates, the slope is again $\tan \phi$. But what about in "unprimed" coordinates? The x' axis itself makes an angle θ with the x-axis. It is tempting to simply add these slopes, obtaining $\tan \phi + \tan \theta$, but this is not correct. Slopes don't add, angles do. The correct answer is that

$$y = x \tan(\theta + \phi), \tag{3.16}$$

so that the slope is given by Equation (3.8).

Remember this discussion when we discuss the Einstein addition formula.

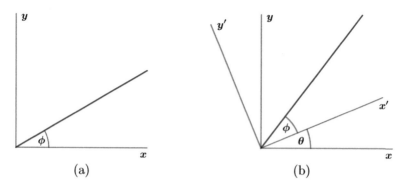

(a)   (b)

FIGURE 3.8. The addition formula for slopes. The given line makes an angle ϕ with (a) the x-axis and (b) the x'-axis.

4

Hyperbola Geometry

In which a two-dimensional non-Euclidean geometry is constructed, which will turn out to be identical with that of special relativity.

4.1 Hyperbolic Trigonometry

The hyperbolic trigonometric functions are usually defined by using the formulas

$$\cosh\alpha = \frac{e^\alpha + e^{-\alpha}}{2}, \tag{4.1}$$

$$\sinh\alpha = \frac{e^\alpha - e^{-\alpha}}{2}, \tag{4.2}$$

$$\tanh\alpha = \frac{\sinh\alpha}{\cosh\alpha}, \tag{4.3}$$

and so on. We give an alternative definition in the next section. The graphs of these hyperbolic functions are shown in Figure 4.1. It is straightforward to verify from these definitions that

$$\cosh^2\alpha - \sinh^2\alpha = 1, \tag{4.4}$$

$$\sinh(\psi + \alpha) = \sinh\psi\cosh\alpha + \cosh\psi\sinh\alpha, \tag{4.5}$$

$$\cosh(\psi + \alpha) = \cosh\psi\cosh\alpha + \sinh\psi\sinh\alpha, \tag{4.6}$$

$$\tanh(\psi + \alpha) = \frac{\tanh\psi + \tanh\alpha}{1 + \tanh\psi\tanh\alpha}, \tag{4.7}$$

$$\frac{d}{d\alpha}\sinh\alpha = \cosh\alpha, \tag{4.8}$$

$$\frac{d}{d\alpha}\cosh\alpha = \sinh\alpha. \tag{4.9}$$

A geometric derivation of these properties is given in Chapter 17.

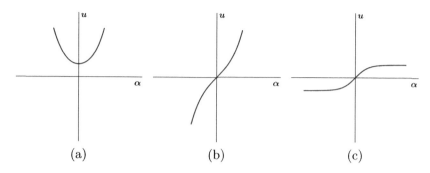

FIGURE 4.1. The graphs of (a) $u = \cosh\alpha$, (b) $u = \sinh\alpha$, and (c) $u = \tanh\alpha$.

The properties of the hyperbolic trigonometric functions look very much like their ordinary trigonometric counterparts (except for signs). This similarity derives from the identities

$$\cosh\alpha \equiv \cos(i\alpha), \qquad (4.10)$$
$$\sinh\alpha \equiv -i\sin(i\alpha). \qquad (4.11)$$

4.2   Distance

We saw in Chapter 3 that Euclidean trigonometry is based on *circles*, sets of points that are a constant distance from the origin. Hyperbola geometry is obtained simply by using a different distance function. Measure the "squared distance" of the point A with coordinates (x, y) from the origin by using the definition

$$\rho^2 = x^2 - y^2. \qquad (4.12)$$

This distance function is the key idea in hyperbola geometry.

4.3   Hyperbola Trigonometry

In hyperbola geometry, "circles" of constant distance from the origin become hyperbolas with $\rho =$ constant. We further restrict ourselves to the branch with $x > 0$. If A is a point on this hyperbola, then we can *define* the hyperbolic angle α between the line from the origin to A and the positive x-axis to be the ratio of the *Lorentzian* length[1] of the arc of the hyperbola between A and the point $(\rho, 0)$ to the "radius" ρ. We could then *define* the hyperbolic trig functions in terms of the coordinates (x, y) of A; that is,

$$\cosh \alpha = \frac{x}{\rho}, \qquad (4.13)$$

$$\sinh \alpha = \frac{y}{\rho}. \qquad (4.14)$$

A little work shows that this definition is equivalent to the one given by Equations (4.1) and (4.2).[2] This construction is shown in Figure 4.2, which also shows another hyperbola, given by $x^2 - y^2 = -\rho^2$. By symmetry, the point B on this hyperbola has coordinates $(x, y) = (\rho \sinh \alpha, \rho \cosh \alpha)$. The importance of this hyperbola will become clear in Chapter 6.

Many of the features of the graphs shown in Figure 4.1 follow immediately from this definition of the hyperbolic trigonometric functions in terms of coordinates along this hyperbola. Since the minimum value of x on this hyperbola is ρ, we must have $\cosh \alpha \geq 1$. As α approaches $\pm\infty$, x approaches ∞ and y approaches $\pm\infty$, which agrees with the asymptotic behavior of the graphs of $\cosh \alpha$ and $\sinh \alpha$, respectively. Finally, since the hyperbola has asymptotes $y = \pm x$, we see that $|\tanh \alpha| < 1$, and that $\tanh \alpha$ must approach ± 1 as α approaches $\pm\infty$.

A unified summary of the hyperbolic trigonometric functions is shown in Figure 4.3; compare Figure 3.3.

So how do we measure the distance between two points? The squared distance defined in (4.12) can be positive, negative, or

---

[1]No, we haven't defined this. In Euclidean geometry, the length of a curve is obtained by integrating ds along the curve, where $ds^2 = dx^2 + dy^2$. In a similar way, the Lorentzian length is obtained by integrating $d\sigma$, where $d\sigma^2 = |dx^2 - dy^2|$.

[2]Use $x^2 - y^2 = \rho^2$ to compute

$$\rho^2 \, d\alpha^2 \equiv d\sigma^2 = dy^2 - dx^2 = \frac{\rho^2 \, dx^2}{x^2 - \rho^2} = \frac{\rho^2 \, dy^2}{y^2 + \rho^2},$$

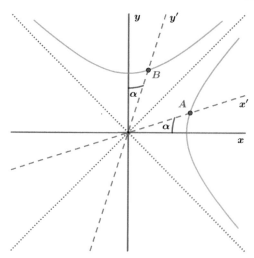

FIGURE 4.2. Defining the hyperbolic trigonometric functions via a hyperbola. Point A has coordinates $(\rho \cosh \alpha, \rho \sinh \alpha)$, and point B has coordinates $(\rho \sinh \alpha, \rho \cosh \alpha)$, where ρ is the *hyperbolic* radius.

zero. We thus adopt the following convention: *Take the square root of the absolute value of the "squared distance."* As shown in Chapter 5, it is important to remember whether the squared distance is positive or negative, but this corresponds directly to whether the distance is "mostly horizontal" or "mostly vertical."

4.4   Triangle Trigonometry

We now recast ordinary triangle trigonometry into hyperbola geometry. Suppose you know $\tanh \alpha = \frac{3}{5}$, and you wish to determine $\cosh \alpha$. One can, of course, do this algebraically by using the identity

$$\cosh^2 \alpha = \frac{1}{1 - \tanh^2 \alpha}. \tag{4.15}$$

then take the square root of either expression and integrate. (The integrals are hard.) Finally, solving for x or y in terms of α yields Equation (4.1) or (4.2), respectively, and the other equation then follows immediately from (4.4).

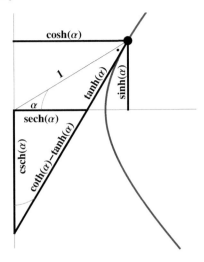

FIGURE 4.3. A unified summary of the hyperbolic trigonometric functions.

But it is easier to draw *any* triangle containing an angle whose hyperbolic tangent is $\frac{3}{5}$. In this case, the obvious choice would be the triangle shown in Figure 4.4, with sides of 3 and 5.

What is $\cosh\alpha$? We first need to work out the length ρ of the hypotenuse. The (hyperbolic) Pythagorean theorem tells us that

$$5^2 - 3^2 = \rho^2, \tag{4.16}$$

so ρ is clearly 4. Take a good look at the 3–4–5 triangle of hyperbola

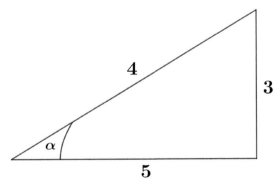

FIGURE 4.4. A hyperbolic right triangle with $\tanh\alpha = \frac{3}{5}$.

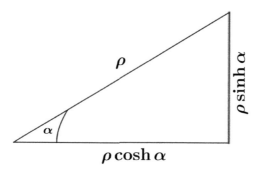

FIGURE 4.5. A hyperbolic right triangle in which the hypotenuse and one angle are known.

geometry shown in Figure 4.4. Now that we know all the sides of the triangle, it is easy to see that $\cosh\alpha = \frac{5}{4}$.

Trigonometry is not merely about the ratios of the sides of triangles; it is also about projections. Another common use of triangle trigonometry is to determine the sides of a triangle given the hypotenuse ρ and one angle α. The answer, of course, is that the sides are $\rho\cosh\alpha$ and $\rho\sinh\alpha$, as shown in Figure 4.5.

4.5   Rotations

By analogy with the Euclidean case, we *define* a hyperbolic rotation through the relations

$$\begin{pmatrix} x \\ y \end{pmatrix} = \begin{pmatrix} \cosh\alpha & \sinh\alpha \\ \sinh\alpha & \cosh\alpha \end{pmatrix} \begin{pmatrix} x' \\ y' \end{pmatrix}. \qquad (4.17)$$

This corresponds to "rotating" both the x- and y-axes into the first quadrant, as shown in Figure 4.2. While this may seem peculiar, it is easily verified that the "distance" is invariant, that is,

$$x^2 - y^2 \equiv x'^2 - y'^2, \qquad (4.18)$$

which follows immediately from the hyperbolic trigonometric identity (4.4).

4.6 Projections

We can ask the same question as we did for Euclidean geometry: Consider a rectangle of width 1 whose sides are parallel to the unprimed axes; how wide is it when measured in the primed coordinates? It turns out that the width of the box in the primed coordinate system is *less than* 1. This is length contraction, to which we will return in Chapter 6, along with time dilation.

4.7 Addition Formulas

What is the slope of the line from the origin to point A in Figure 4.2? The equation of this line, the y'-axis, is

$$x = y \tanh \alpha. \tag{4.19}$$

Consider now a line with equation

$$x' = y' \tanh \psi. \tag{4.20}$$

What is its (unprimed) slope? Again, slopes don't add, but hyperbolic angles do; therefore,

$$x = y \tanh(\psi + \alpha), \tag{4.21}$$

which can be expressed in terms of the slopes $\tanh \psi$ and $\tanh \alpha$ by using Equation (4.7). As discussed in more detail in Chapter 6, this is the Einstein addition formula.

4.8 Combining Circle and Hyperbola Trigonometry

Can circle and hyperbola trigonometry ever be used at the same time? Although at first sight unlikely, there is indeed some overlap between these two notions of trigonometry.

For example, consider the identity

$$\text{sech}^2\alpha + \tanh^2\alpha = \frac{1 + \sinh^2\alpha}{\cosh^2\alpha} = 1 \qquad (4.22)$$

in hyperbola trigonometry, which resembles the fundamental identity in circle trigonometry, namely $\cos^2\theta + \sin^2\theta = 1$. So what happens if we set corresponding trigonmetric functions equal? If we implicitly relate α and θ via

$$\text{sech}\alpha = \cos\theta, \qquad (4.23)$$
$$\tanh\alpha = \sin\theta, \qquad (4.24)$$

we have immediately also that

$$\sinh\alpha = \tan\theta, \qquad (4.25)$$
$$\cosh\alpha = \frac{1}{\cos\theta}. \qquad (4.26)$$

Converting to exponential form, we have

$$\sin\theta = \tanh\alpha = \frac{\sinh\alpha}{\cosh\alpha} = \frac{e^\alpha - e^{-\alpha}}{e^\alpha + e^{-\alpha}} \qquad (4.27)$$

which we can solve for α, yielding

$$\alpha = \frac{1}{2}\ln\frac{1 + \sin\theta}{1 - \sin\theta}. \qquad (4.28)$$

However, care must be taken with the domain of θ, since the right-hand side is only continuously defined for $-\frac{\pi}{2} < \theta < \frac{\pi}{2}$ (say). Nonetheless, on this restricted domain it is indeed possible to use these trigonometric functions interchangeably. In support of this unexpected conclusion we can check that the ranges actually match. Which they do, since both $\tanh\alpha$ and $\text{sech}\alpha = \frac{1}{\cosh\alpha}$ have magnitude less than 1, and $\tan\theta$ takes on all real values.

But these domain restrictions are key, as circle trigonometric functions are periodic, while hyperbola functions are not. So although it is, surprisingly, possible to relate them in some cases, in most applications it is readily apparent that one is a better match than the other.

5

The Geometry of Special Relativity

In which it is shown that special relativity is just hyperbola geometry.

5.1 The Surveyors

A brilliant aid in understanding special relativity is the *surveyors' parable* introduced by Taylor and Wheeler [Taylor and Wheeler 63, Taylor and Wheeler 92]. Suppose a town has daytime surveyors, who determine north and east with a compass, and nighttime surveyors, who use the North Star. These notions differ of course, since magnetic north is not the direction to the North Pole. Suppose, further, that both groups measure north/south distances in miles and east/west distances in meters, with both being measured from the town center. How does one go about comparing the measurements of the two groups?

With our knowledge of Euclidean geometry, we see how to do this: convert miles to meters (or vice versa). Distances computed with the Pythagorean theorem do not depend on which group does the surveying. Finally, it is easily seen that "daytime" coordinates can be obtained from "nighttime" coordinates by a simple rotation. The geometry of this situation is therefore described by Figure 3.6, in which the x and y directions correspond to geographic east and north, respectively, and the x' and y' directions correspond to *magnetic* east and north, respectively. If the surveyors measure x and x' in meters, and y and y' in miles, and if they do not understand how to convert between the two, communication between the two groups will not be easy.

The moral of this parable is therefore:

1. *The same units should be used for all distances.*

2. *The (squared) distance is invariant.*

3. *Different frames are related by rotations.*

Applying this lesson to special relativity, we should measure both time and space in the same units. How do we measure distance in seconds? That's easy: One second of distance is the distance traveled (in vacuum) by light in 1 second.[1] This has the effect of setting $c = 1$, since the number of seconds (of distance) traveled by light in 1 second (of time) is precisely 1.

Of course, it is also possible to measure time in meters: 1 meter of time is the time it takes for light (in vacuum) to travel 1 meter. Again, this has the effect of setting $c = 1$. To convert seconds to meters, simply multiply by c. For example, since $c = 3 \times 10^8$ m/s, 1 second of distance is 3×10^8 meters.

5.2   Spacetime Diagrams

A *spacetime diagram* is a diagram showing both where things are and when they were there. For example, Figure 5.1(a) represents somebody sitting still (but getting older!), and Figure 5.1(b) represents somebody moving to the right at constant speed. Such trajectories are called *worldlines*.[2] In relativity, the convention is to have the t-axis be vertical, and the x-axis be horizontal.

The surveyors' parable tells us that we should measure space and time in the same units, not one in meters and the other in seconds, but both in either meters or seconds. We choose meters; this amounts to using ct to measure time rather than t. In either case, light beams play a special role in spacetime diagrams because they are drawn at 45°.

---

[1]A similar unit of distance is the *lightyear*, namely the distance traveled by light in 1 year, which would here be called simply a *year* of distance.

[2]One early French railroad timetable actually shows the worldlines of each train.

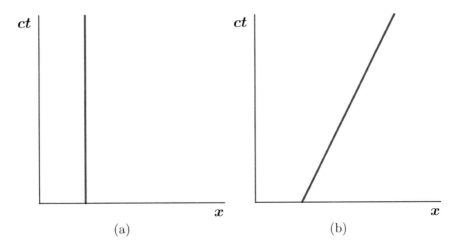

FIGURE 5.1. Spacetime diagrams representing (a) standing still and (b) moving to the right at constant speed.

We will often implement these conventions without further comment, namely setting $c = 1$ and assuming that the vertical axis denotes time. We will, however, usually label this axis "ct" as a reminder.

One fundamental geometric difference between circle trigonometry and hyperbola trigonometry is the presence of asymptotes in the hyperbolic case. These asymptotes have physical significance. Recall that inverse slopes in a spacetime diagram correspond to speeds; the asymptotes correspond to the existence of a "special" speed, namely, the speed of light.

We explore the properties of spacetime diagrams throughout the remainder of this chapter; the results are summarized in Section 6.1.

5.3   Lorentz Transformations

We now relate Lorentz transformations, based on the physical postulates of special relativity, to hyperbola geometry. The Lorentz transformation between a frame (x, t) at rest and a frame (x', t')

moving to the right at speed v was derived in Chapter 2. The transformation from the moving frame to the frame at rest is given by

$$x = \gamma \left(x' + vt' \right), \tag{5.1}$$

$$t = \gamma \left(t' + \frac{v}{c^2} x' \right), \tag{5.2}$$

where, as before,

$$\gamma = \frac{1}{\sqrt{1 - \frac{v^2}{c^2}}}. \tag{5.3}$$

The key to converting this description to hyperbola geometry is to measure space and time in the same units by replacing t by ct. The transformation from the moving frame, which we now denote (x', ct'), to the frame at rest, now denoted (x, ct), is given by

$$x = \gamma \left(x' + \frac{v}{c} ct' \right), \tag{5.4}$$

$$ct = \gamma \left(ct' + \frac{v}{c} x' \right), \tag{5.5}$$

which makes the symmetry between these equations much more obvious.

We can simplify things still further by introducing the *rapidity* α via[3]

$$\frac{v}{c} = \tanh \alpha. \tag{5.6}$$

By inserting Equation (5.6) into the expression for γ, (5.3), we obtain the identities

$$\gamma = \frac{1}{\sqrt{1 - \tanh^2 \alpha}} = \sqrt{\frac{\cosh^2 \alpha}{\cosh^2 \alpha - \sinh^2 \alpha}} = \cosh \alpha \tag{5.7}$$

and

$$\frac{v}{c} \gamma = \tanh \alpha \cosh \alpha = \sinh \alpha \tag{5.8}$$

By then inserting identities (5.7) and (5.8) into the Lorentz transformations (5.4) and (5.5), we obtain the remarkably simple form

$$x = x' \cosh \alpha + ct' \sinh \alpha, \tag{5.9}$$

$$ct = x' \sinh \alpha + ct' \cosh \alpha \tag{5.10}$$

---

[3]Warning: Some authors use α for $\frac{v}{c}$, not the rapidity.

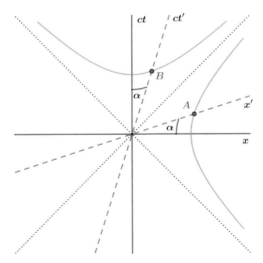

FIGURE 5.2. The Lorentz transformation between an observer at rest and an observer moving at speed $\frac{v}{c} = \tanh\alpha$ is shown as a hyperbolic rotation. Point B has coordinates $(\rho\sinh\alpha, \rho\cosh\alpha)$. Compare Figure 4.2.

which, in matrix form, is

$$\begin{pmatrix} x \\ ct \end{pmatrix} = \begin{pmatrix} \cosh\alpha & \sinh\alpha \\ \sinh\alpha & \cosh\alpha \end{pmatrix} \begin{pmatrix} x' \\ ct' \end{pmatrix}, \tag{5.11}$$

which is just Equation (4.17), with $y = ct$.

Thus, Lorentz transformations are just hyperbolic rotations. As noted in Chapter 4, the invariance of the interval follows immediately from the fundamental hyperbolic trigonometric identity (4.4). This invariance now takes the form

$$x^2 - c^2t^2 \equiv x'^2 - c^2t'^2, \tag{5.12}$$

as shown in Figure 5.2, which is nearly identical to Figure 4.2. The physics of Lorentz transformations and the Lorentzian geometry of hyperbolas are therefore one and the same; hyperbola geometry is the geometry of special relativity.

5.4   Space and Time

We now return to the peculiar fact that the squared distance between two points can be positive, negative, or zero. The sign is positive for horizontal distances and negative for vertical distances. These directions correspond to the coordinates x and t, and measure space and time, respectively—as seen by the given observer. But *any* observer's space axis must intersect the hyperbola $x^2 - c^2t^2 = \rho^2$ somewhere, and hence corresponds to positive squared distance. Such directions have more space than time, and are called *spacelike*. Similarly, any observer's time axis intersects the hyperbola $x^2 - c^2t^2 = -\rho^2$, corresponding to negative squared distance; such directions are called *timelike*.

What about diagonal lines at a Euclidean angle of 45°? These lines correspond to a squared distance of zero—and to moving at the speed of light. All observers agree about these directions, which are called *lightlike*. In hyperbola geometry, there are thus preferred directions of "length zero." Indeed, this is the geometric realization of the idea that the speed of light is the same for all observers.

It is important to realize that *every* spacelike direction corresponds to the space axis for *some* observer. Events separated by a spacelike line occur at the same time for that observer—and the square root of the squared distance is just the distance between the events as seen by that observer. Similarly, events separated by a timelike line occur at the same place for some observer, and the square root of the absolute value of the squared distance is just the time that elapses between the events as seen by that observer.

In contrast, events separated by a timelike line do not occur simultaneously for *any* observer. We can thus divide the space-time diagram into causal regions as follows: those points connected to the origin by spacelike lines occur "now" for some observer, whereas those points connected to the origin by timelike lines occur unambiguously in the future or the past. This relationship is shown in Figure 5.3.[4]

---

[4]With two or more spatial dimensions, the lightlike directions would form a surface called the *light cone*, and the regions labeled "now" would be connected.

FIGURE 5.3. The causal relationship between points in spacetime and the origin.

To be able to make sense of cause and effect, only events in our past can influence us, and we can influence only events in our future. Put differently, if information could travel faster than the speed of light, then different observers would no longer be able to agree on cause and effect.

5.5   The Geometry of Lorentz Transformations

In Section 5.3, we used algebra to show that the Lorentz transformations derived on physical grounds in Chapter 2 are precisely the hyperbolic rotations discussed in Chapter 4. Here, we use geometry to argue the other way around, namely that hyperbolic rotations are in fact Lorentz transformations.

The relationship between two inertial frames is shown in Figure 5.4. The vertical axis represents the worldline of an observer at rest, and the horizontal axis represents a line of simultaneity for such an observer. This inertial observer describes events using coordinates (x, t). Similarly, the slanted timelike axis represents the worldline of an observer moving to the right with speed $v = \tanh\alpha$, and the slanted spacelike axis represents a line of simultaneity for such an observer. This observer describes events

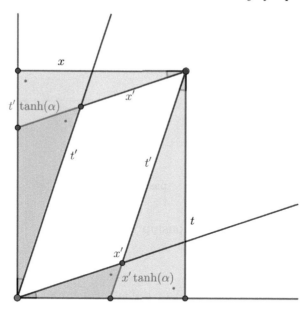

FIGURE 5.4. The geometric relationship between inertial reference frames. Hyperbolic projections first in the small triangles (near the origin), then in the large triangles (away from the origin) yield the standard formulas for the Lorentz transformation, expressed as a hyperbolic rotation. Units have been chosen such that $c = 1$.

using coordinates (x', t'). Several instances of the hyperbolic angle α are shown; (some) right angles are indicated by dots.

The geometric interpretation of the coordinates (x, t) and (x', t') of an arbitrary point (upper right) as "distances" to the appropriate axis are shown in the figure. Starting with the small right triangle at the bottom (with tip at the origin), the adjacent leg has length x', so the opposite leg has length $x' \tanh \alpha$. Looking now at the large right triangle on the right (with tip at the arbitrary point), the hypotenuse has length $t' + x' \tanh \alpha$, whereas the adjacent leg has length t. Thus,

$$t = (t' + x' \tanh \alpha) \cosh \alpha = t' \cosh \alpha + x' \sinh \alpha. \qquad (5.13)$$

Exactly the same construction applied to the other two triangles yields the same result with t and x switched, namely

$$x = (x' + t' \tanh \alpha) \cosh \alpha = x' \cosh \alpha + t' \sinh \alpha. \qquad (5.14)$$

Equations (5.13) and (5.14) are the Lorentz transformation taking the coordinates (x', t') of the moving observer into the coordinates (x, t) of the observer at rest, expressed as a hyperbolic rotation.

There are several ways to derive the inverse transformation, from the observer at rest to the moving observer. By symmetry, the only thing that changes is the sign of the relative velocity, so that α is replaced by $-\alpha$. Thus,

$$t' = t \cosh \alpha - x \sinh \alpha,$$
$$x' = x \cosh \alpha - t \sinh \alpha, \tag{5.15}$$

as can also be easily verified algebraically, using the fundamental identity of hyperbolic geometry, namely

$$\cosh^2 \alpha - \sinh^2 \alpha = 1. \tag{5.16}$$

It is also straightforward to adapt Figure 5.4 to the case where $\alpha < 0$, then swap the primed and unprimed variables. Finally, and perhaps most interestingly, it is possible to use the figure as drawn, with a different choice of triangles, to derive (5.15) directly.

5.6   Dot Product

In Euclidean geometry, (squared) distances can be described by taking the dot product of a vector with itself. By denoting the unit vectors in the x and y directions by $\hat{x}$ and $\hat{y}$, respectively, we can write the vector from the origin to the point (x, y) as

$$\vec{r} = x\,\hat{x} + y\,\hat{y}, \tag{5.17}$$

whose squared length is

$$|\vec{r}|^2 = \vec{r} \cdot \vec{r} = x^2 + y^2. \tag{5.18}$$

The generalization to hyperbola geometry is straightforward. We denote the unit vectors in the t and x directions by $\hat{t}$ and $\hat{x}$.[5]

---

[5] Unit vectors are dimensionless. It is neither necessary nor desirable to include a factor of c in the definition of $\hat{t}$.

Then the Lorentzian dot product can be defined by the requirement that this be an orthonormal basis in the sense that

$$\hat{x} \cdot \hat{x} = 1, \tag{5.19}$$

$$\hat{t} \cdot \hat{t} = -1, \tag{5.20}$$

$$\hat{x} \cdot \hat{t} = 0. \tag{5.21}$$

Any point $(x,\ ct)$ in spacetime can thus be identified with the vector

$$\vec{r} = x\,\hat{x} + ct\,\hat{t} \tag{5.22}$$

from the origin to that point, whose squared length is just the squared distance from the origin, namely

$$|\vec{r}|^2 = \vec{r} \cdot \vec{r} = x^2 - c^2 t^2. \tag{5.23}$$

One of the fundamental properties of the Euclidean dot product is that

$$\vec{u} \cdot \vec{v} = |\vec{u}||\vec{v}| \cos\theta, \tag{5.24}$$

where θ is the smallest angle between the directions of $\vec{u}$ and $\vec{v}$. This relationship between the dot product and projections of one vector along another can in fact be used to *define* the dot product. But what happens in hyperbola geometry?

First of all, the dot product can be used to define right angles: Two vectors $\vec{u}$ and $\vec{v}$ are said to be *orthogonal* (or *perpendicular*) precisely when their dot product is zero:

$$\vec{u} \perp \vec{v} \iff \vec{u} \cdot \vec{v} = 0. \tag{5.25}$$

We adopt this definition unchanged in hyperbola geometry.

When are $\vec{u}$ and $\vec{v}$ perpendicular? We assume first that $\vec{u}$ is spacelike. We can assume without loss of generality that $\vec{u}$ is a unit vector with positive x-component, in which case it takes the form

$$\vec{u} = \cosh\alpha\,\hat{x} + \sinh\alpha\,\hat{t}. \tag{5.26}$$

Then one vector perpendicular to $\vec{u}$ would be

$$\vec{v} = \sinh\alpha\,\hat{x} + \cosh\alpha\,\hat{t} \tag{5.27}$$

and it is easy to check that all other vectors perpendicular to $\vec{u}$ are multiples of this one. Note that $\vec{v}$ is timelike. Had we assumed instead that $\vec{u}$ were timelike, we would merely have interchanged the roles of $\vec{u}$ and $\vec{v}$.

Furthermore, $\vec{u}$ and $\vec{v}$ are just the space and time axes, respectively, of an observer moving with speed $\frac{v}{c} = \tanh\alpha$. So orthogonal directions correspond precisely to the coordinate axes of some observer.

What if $\vec{u}$ is lightlike? It is a peculiarity of Lorentzian hyperbola geometry that there are nonzero vectors of length zero. But since the dot product of a vector with itself gives its squared length, having length zero means that lightlike vectors are perpendicular to themselves.

We can finally define the *length* of a vector $\vec{v}$ by

$$|\vec{v}| = \sqrt{|\vec{v} \cdot \vec{v}|}. \tag{5.28}$$

If $\vec{v}$ is spacelike, we can write

$$\vec{v} = \pm|\vec{v}|(\cosh\alpha\,\hat{\boldsymbol{x}} + \sinh\alpha\,\hat{\boldsymbol{t}}), \tag{5.29}$$

whereas if $\vec{v}$ is timelike, we can write

$$\vec{v} = \pm|\vec{v}|(\sinh\alpha\,\hat{\boldsymbol{x}} + \cosh\alpha\,\hat{\boldsymbol{t}}). \tag{5.30}$$

If $\vec{v}$ is lightlike, however, $|\vec{v}| = 0$, so no such expression exists. Rather,

$$\vec{v} = a\,(\hat{\boldsymbol{x}} \pm \hat{\boldsymbol{t}}) \tag{5.31}$$

for some constant a.

The above argument shows that timelike vectors can only be perpendicular to spacelike vectors, and vice versa. We also say in this case that the vectors form a *right angle*. Recall that hyperbolic angles are defined along hyperbolas of the form $x^2 - c^2t^2 = \rho^2$, and hence exist (as originally defined) only between spacelike directions. It is straightforward to extend this definition to timelike directions by using the hyperbola $x^2 - c^2t^2 = -\rho^2$; this was implicitly done in Figure 4.2. But there is no hyperbola relating timelike directions to spacelike ones. Thus, a "right angle" isn't an angle at all!

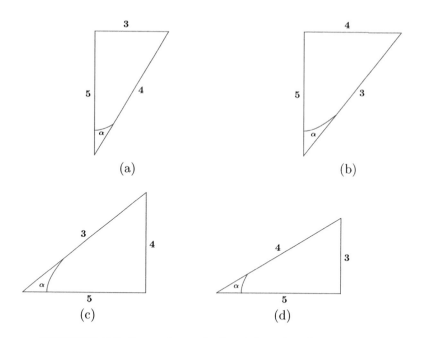

FIGURE 5.5. Some hyperbolic right triangles.

A *right triangle* is one that contains a right angle. By the above discussion, one of the legs of such a triangle must be spacelike and the other timelike. Consider first the case where the hypotenuse is either spacelike or timelike. The only hyperbolic angle in such a triangle is the one between the hypotenuse and the leg of the same type, that is, between the two timelike sides if the hypotenuse is timelike and between the two spacelike sides if the hypotenuse is spacelike. Several such hyperbolic right triangles are shown in Figures 5.5(a)–(d). It is also possible for the hypotenuse to be null, as shown in Figure 5.6; such triangles do not have any hyperbolic angles. Finally, as seen in Figures 5.6(b) and 5.6(c), right angles do not always look like right angles. The reader is encouraged to combine these ideas, and attempt to draw a 3–4–5 triangle, none of whose sides are horizontal or vertical.

What happens if we take the dot product between two spacelike vectors? We can assume without loss of generality that one vector

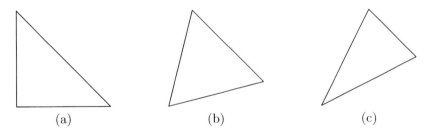

(a)   (b)   (c)

FIGURE 5.6. More hyperbolic right triangles. The right angle is on the left!

is parallel to the (positive) x-axis, in which case we have

$$\vec{u} = |\vec{u}|\,\hat{x}, \tag{5.32}$$
$$\vec{v} = \pm|\vec{v}|(\cosh\alpha\,\hat{x} + \sinh\alpha\,\hat{t}), \tag{5.33}$$

so that the dot product satisfies

$$\vec{u} \cdot \vec{v} = \pm|\vec{u}||\vec{v}|\cosh\alpha. \tag{5.34}$$

This result does indeed match the standard result from Euclidean geometry, since the angle α must be geometrically smaller than a right angle; in Euclidean terms, it is the reference angle between the vectors, which is always acute.

What happens if both vectors are timelike? The above argument still works, except that the roles of $\hat{x}$ and $\hat{t}$ must be interchanged, resulting in

$$\vec{u} \cdot \vec{v} = \mp|\vec{u}||\vec{v}|\cosh\alpha, \tag{5.35}$$

where the sign is now positive if both vectors point toward the future (or past), and negative otherwise. In both the spacelike and timelike cases, $|\vec{v}|\cosh\alpha$ is the projection of $\vec{v}$ along $\vec{u}$; see Figure 5.7.

What happens if we take the dot product between a timelike vector and a spacelike vector? We can again assume without loss of generality that the spacelike vector is parallel to the (positive) x-axis, so that

$$\vec{u} = |\vec{u}|\,\hat{x}, \tag{5.36}$$
$$\vec{v} = |\vec{v}|(\sinh\alpha\,\hat{x} + \cosh\alpha\,\hat{t}). \tag{5.37}$$

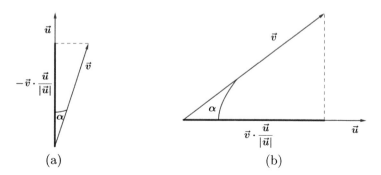

(a)   (b)

FIGURE 5.7. The hyperbolic projection of $\vec{v}$ onto $\vec{u}$ for (a) two timelike vectors, or (b) two spacelike vectors.

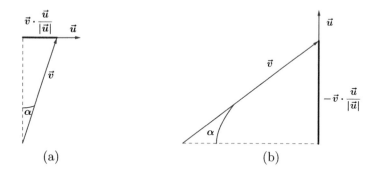

(a)   (b)

FIGURE 5.8. The hyperbolic projection of $\vec{v}$ onto $\vec{u}$ when one is timelike and the other spacelike. The orientation of $\vec{u}$ in each case is the opposite of that shown in Figure 5.7, but $\vec{v}$ is unchanged.

The dot product now satisfies

$$\vec{u} \cdot \vec{v} = |\vec{u}||\vec{v}| \sinh \alpha. \tag{5.38}$$

At first sight, this is something new. But note from Figure 5.8(a) that $\vec{v} \sinh \alpha$ is just the projection of $\vec{v}$ along $\vec{u}$! The new feature here is that we cannot define the angle between a timelike direction and a spacelike direction. The only angle in the triangle that is defined is the one shown.[6]

---

[6]Alternatively, we could have assumed that the timelike vector was parallel to the (positive) t-axis, resulting in Figure 5.8(b). The conclusion is the same, although now it represents the projection of $\vec{u}$ along $\vec{v}$.

6

Applications

In which we learn some of the consequences of special relativity.

6.1 Drawing Spacetime Diagrams

We begin by summarizing the rules for drawing spacetime diagrams.

- Points in spacetime are called *events*.

- Lines with slope $m = \pm 1$ represent beams of light.

- Vertical lines represent the worldline of an object at rest.

- Horizontal lines represent snapshots of constant time, that is, events that are simultaneous (in the given reference frame).

- Lines with slope $|m| > 1$ (called *timelike*) represent the worldlines of observers moving at constant speed.

- The speed of such observers is given by $c \tanh \alpha$, where α is the (hyperbolic) angle between the worldline and a *vertical* line.

- The distance between two events along a timelike line is just the time between them as measured by the moving observer.

- Lines with slope $|m| < 1$ (called *spacelike*) represent lines of simultaneity as seen by an observer moving at constant speed.

- The speed of this observer is given by $c \tanh \alpha$, where α is the (hyperbolic) angle between the line of simultaneity and a *horizontal* line.

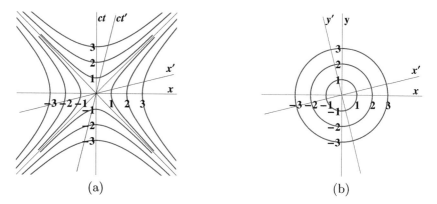

(a)   (b)

FIGURE 6.1. The first diagram shows the use of hyperbolas to calibrate spacetime diagrams. Several hyperbolas are shown, each corresponding to constant "distance" from the origin. The analogous calibration diagram in circle geometry is shown in the second diagram. In both cases, the calibration curves can be used to compare distances measured in one frame to those measured in another.

- The distance between two events along a spacelike line is just the distance between them as measured by this observer.

- Hyperbolas centered at the origin represent events at constant "distance" from the origin; such hyperbolas can be used to calibrate the scales along any line, as illustrated in Figure 6.1.

- Lines are orthogonal if they have reciprocal slopes.

- Triangle trigonometry can be used to relate measurements in different reference frames. *(Be careful: A right triangle contains only one angle.)*

6.2   Addition of Velocities

What does the rapidity α represent? Consider an observer moving at speed v to the right. This observer's world line intersects the hyperbola

$$c^2 t^2 - x^2 = \rho^2 \qquad (ct > 0) \qquad\qquad (6.1)$$

at the point B with coordinates $(\rho \sinh \alpha, \rho \cosh \alpha)$; this line therefore has "slope"[1]

$$\frac{v}{c} = \tanh \alpha \qquad (6.2)$$

which is the same as Equation (5.6). Thus, α is nothing more than the *hyperbolic angle* between the *ct*-axis and the worldline of a moving object. As discussed in Chapter 4, $\rho\alpha$ is precisely the distance from the axis as measured along the hyperbola (in hyperbola geometry). This relationship was illustrated in Figures 4.2 and 5.2.

Consider, therefore, an object moving at speed u relative to an observer moving at speed v. Their rapidities are given by, respectively,

$$\frac{u}{c} = \tanh \psi, \qquad (6.3)$$

$$\frac{v}{c} = \tanh \alpha. \qquad (6.4)$$

To determine the resulting speed with respect to an observer at rest, we simply add the *rapidities*! One way to think of this process is that we are adding the arc lengths along the hyperbola. Another is that we are following a hyperbolic rotation through a hyperbolic angle α (to get to the moving observer's frame) with a rotation through an angle ψ, as shown in Figure 6.2. In any case, the resulting speed w is given by

$$\frac{w}{c} = \tanh(\psi + \alpha) = \frac{\tanh \psi + \tanh \alpha}{1 + \tanh \psi \tanh \alpha} = \frac{\frac{u}{c} + \frac{v}{c}}{1 + \frac{uv}{c^2}}, \qquad (6.5)$$

which is—finally—precisely the Einstein addition formula.

6.3   Length Contraction

We now return to the question of how "wide" things are. We consider first a meter stick at rest. In spacetime, the stick "moves"

---

[1]It is not obvious whether slope should be defined by $\frac{\Delta x}{c\Delta t}$ or by the reciprocal of this expression. This situation is further complicated by the fact that both (x, ct) and (ct, x) are commonly used to denote the coordinates of the point A.

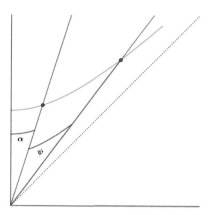

FIGURE 6.2. The geometry of the Einstein addition formula.

vertically; that is, it ages. This situation is shown in Figure 6.3(a), where the horizontal lines show the meter stick at various times (according to an observer at rest). How "wide" is the *worldsheet* of the stick? The observer at rest, of course, measures the length of the stick by locating both ends *at the same time* and measuring the distance between them. At $t = 0$, these events correspond to the two heavy dots in the sketch, one at the origin and the other on the unit hyperbola. But *all* points on the unit hyperbola are at an interval of 1 meter from the origin. The observer at rest therefore concludes, unsurprisingly, that the meter stick is 1 meter long.

How long does a moving observer think the stick is? This length is just the width of the worldsheet *as measured by the moving observer.* This observer follows the same procedure, by locating both ends of the stick *at the same time*, and measuring the distance between them. But time now corresponds to t', not t. At $t' = 0$, this measurement corresponds to the heavy line in Figure 6.3(a), Since this line fails to reach the unit hyperbola, it is clear that the moving observer measures the length of a stationary meter stick to be less than 1 meter. This effect is length contraction.

To determine the exact value measured by the moving observer, it is necessary only to note that the triangle shown in Figure 6.3(a) is a hyperbolic right triangle, with the heavy line indicating the hypotenuse. Simple hyperbolic trigonometry shows immediately

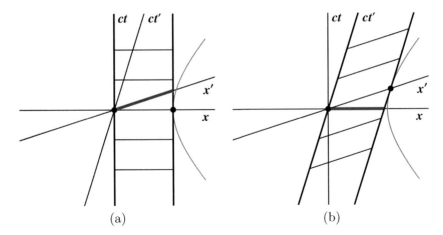

FIGURE 6.3. Length contraction as a hyperbolic projection: (a) object at rest and observer in motion; (b) object in motion and observer at rest.

that the hypotenuse of a right triangle with angle α and adjacent side of length 1 is given by $1/\cosh\alpha$.

For those who prefer an algebraic derivation, we compute the intersection of the line $x = 1$ (the right-hand edge of the meter stick) with the line $t' = 0$ (the x'-axis), or equivalently, $ct = x\tanh\alpha$, to find that

$$ct = \tanh\alpha, \tag{6.6}$$

so that x' is just the interval from this point to the origin, which is

$$x' = \sqrt{x^2 - c^2t^2} = \sqrt{1 - \tanh^2\alpha} = \frac{1}{\cosh\alpha}. \tag{6.7}$$

What if the stick is moving and the observer is at rest? This situation is shown in Figure 6.3(b). The worldsheet now corresponds to a "rotated rectangle," indicated by the parallelograms in the sketch. The fact that the meter stick is 1 meter long in the moving frame is shown by the distance between the two heavy dots (along $t' = 0$), and the measurement by the observer at rest is indicated by the heavy line (along $t = 0$). Again, it is clear that the stick appears to have shrunk, since the heavy line fails to reach

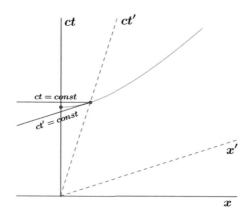

FIGURE 6.4. Time dilation as a hyperbolic projection.

the unit hyperbola. To determine the exact value measured by the observer at rest, proceed as before using the hyperbolic right triangle shown in Figure 6.3(b), which is, however, congruent to the triangle shown in Figure 6.3(a), so that the result is the same.

Thus, a moving object appears shorter by a factor of $1/\cosh\alpha$. It doesn't matter whether it is the stick or the observer that is moving; all that matters is their relative motion.

6.4   Time Dilation

We now investigate moving clocks. Consider first the smaller dot in Figure 6.4. This dot corresponds to $ct = 1$ (and $x = 0$), as evidenced by the fact that this point is on the other unit hyperbola, as shown. Similarly, the larger dot, lying on the same hyperbola, corresponds to $ct' = 1$ (and $x' = 0$). The horizontal line emanating from this dot gives the value of ct there, which is clearly greater than 1, and which represents the time according to the observer at rest when the moving clock says 1. According to the observer at rest, the moving clock therefore runs slow. But now consider the diagonal line emanating from the larger dot. At all points along this line, $ct' = 1$. In particular, at the smaller dot we must have $ct' > 1$. Thus, the time according to the moving observer when

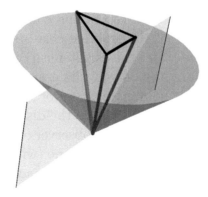

FIGURE 6.5. The three-dimensional spacetime diagram for a bouncing beam of light on a moving train. The beam of light is along the light cone and is projected into both the xy-plane (top) and the xt-plane (front).

the clock at rest says 1 (at the smaller dot) must be greater than 1; the moving observer therefore concludes the clock at rest runs slow.

There is no contradiction here; we must simply be careful to ask the right question. In each case, observing a clock in another frame of reference corresponds to a projection. In each case, a clock in relative motion to the observer appears to run slow.

To determine the exact value measured by the moving observer, we return to the bouncing beam of light considered in Chapter 2. Figure 2.5 showed the relationships between the various *distances*, but we would now like to draw a spacetime diagram for this scenario. However, since the beam of light is moving in both the horizontal (x) and vertical (y) directions, we need a *three-dimensional* spacetime diagram, as shown in Figure 6.5, and discussed in more detail in Section 14.4.

In Figure 6.5, the vertical line represents an observer at rest on the platform, and the mostly vertical line on the right represents the motion of the flashlight on the floor of the moving train. The beam of light is the line in the back of the diagram, along the light cone, and moving in both the x and y directions.

By projecting Figure 6.5 into a horizontal plane (with $t =$ constant), as shown at the top of the cone, we recover precisely Figure 2.5. The base of the triangle corresponds to the distance $v \Delta t$ traveled by the train, the hypotenuse corresponds to the distance $c \Delta t$ traveled by the beam of light according to the observer on the platform, and the remaining leg of the triangle corresponds to the height of the train, that is, to the distance $c \Delta t'$ traveled by the beam of light according to the observer on the train. Projecting Figure 6.5 instead into the vertical plane (with $y = 0$), as shown in the middle of the cone, one obtains the spacetime diagram shown in Figure 6.6.

Remarkably, the edges of the two triangles formed by the horizontal and vertical projections have the same lengths. To see this, we note first of all that they share one edge. But each of the remaining pairs of edges form the legs of a right triangle whose hypotenuse is the beam of light. The speed of light is 1 (in appropriate units), so in each case the legs must have the same length. This argument justifies the labeling used in Figure 6.6, and can in fact be used to *derive* the hyperbolic Pythagorean theorem, which says here that

$$(c \Delta t')^2 = (c \Delta t)^2 - (v \Delta t)^2, \tag{6.8}$$

or equivalently, that

$$\Delta t' = \frac{\Delta t}{\cosh \alpha}, \tag{6.9}$$

where $\tanh \alpha = v/c$ is the speed of the train. Moving clocks therefore run slow, by a factor of precisely $\cosh \alpha$.

It is worth comparing Figures 2.5 and 6.6, using Figure 6.5 as needed. These two diagrams are almost the same, but two of the edges appear to have been interchanged. In Figure 2.5, the vertical *distance* traveled by the beam of light was $c\Delta t'$; in Figure 6.6, the *time* taken by the beam to reach the top of the train is $c\Delta t'$ (in appropriate units), namely, the length of the projected worldline of the beam of light, which is the hypotenuse of the triangle. As suggested by this brief discussion, these two figures can indeed be unified into a single, three-dimensional spacetime diagram; see Figure 14.5 and the discussion in Section 14.4.

FIGURE 6.6. The spacetime Pythagorean theorem for a bouncing light beam on a moving train.

6.5 Doppler Shift

The frequency f of a beam of light is related to its wavelength λ by the formula

$$f\lambda = c. \tag{6.10}$$

How do these quantities depend on the observer?

Let's consider an inertial observer moving to the right in the laboratory frame and carrying a flashlight that is pointing to the left; see Figure 6.7. Then the moving observer, using coordinates (t',x'), is traveling along a path of the form $x' = x_1' = $ constant. Suppose the moving observer turns on the flashlight (at time t_1') just long enough to emit one complete wavelength of light, and that this takes time dt'. Then the moving observer "sees" a wavelength

$$\lambda' = c\,dt'. \tag{6.11}$$

According to the observer in the lab, the flashlight was turned on at the event (t_1, x_1) and turned off dt_1 seconds later, during which time the moving observer moved a distance dx_1 meters to the right. But when was the light received, at, say, $x = 0$?

We let $(t_0, 0)$ denote the first reception of light by a lab observer at $x = 0$, and suppose this observer sees the light stay on for dt_0

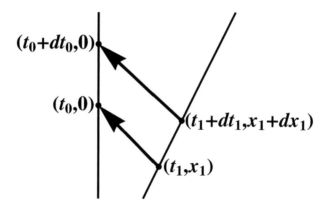

FIGURE 6.7. The Doppler effect. An observer moving to the right emits a pulse of light to the left, which is later seen by a stationary observer. The wavelengths measured by the two observers differ, causing a *Doppler shift* in the frequency. The coordinates of each event are given with respect to the stationary observer.

seconds. Since light travels at the speed of light, we have

$$c(t_0 - t_1) = x_1 \tag{6.12}$$

and

$$c[(t_0 + dt_0) - (t_1 + dt_1)] = x_1 + dx_1, \tag{6.13}$$

from which it follows that

$$c(dt_0 - dt_1) = dx_1, \tag{6.14}$$

so that

$$
\begin{aligned}
c\,dt_0 &= dx_1 + c\,dt_1 \\
&= (dx_1' \cosh\alpha + c\,dt_1' \sinh\alpha) + (c\,dt_1' \cosh\alpha + dx_1' \sinh\alpha) \\
&= (\cosh\alpha + \sinh\alpha)\,c\,dt_1',
\end{aligned}
\tag{6.15}
$$

since $dx_1' = 0$. But the wavelength as seen in the lab is

$$\lambda = c\,dt_0, \tag{6.16}$$

so that

$$\frac{\lambda}{\lambda'} = \frac{dt_0}{dt'_1} = \cosh\alpha + \sinh\alpha$$

$$= \cosh\alpha\,(1 + \tanh\alpha) = \gamma\left(1 + \frac{v}{c}\right) = \sqrt{\frac{1 + \frac{v}{c}}{1 - \frac{v}{c}}}. \qquad (6.17)$$

The frequencies transform inversely, that is,

$$\frac{f'}{f} = \sqrt{\frac{1 + \frac{v}{c}}{1 - \frac{v}{c}}}. \qquad (6.18)$$

7

Problems I

7.1 Warmup

1. *Three events have the following lab coordinates (x, ct):*

$$A = (1, 2), \qquad B = (4, 7), \qquad C = (6, 5), \qquad (7.1)$$

where the units of length are arbitrary.

(a) *Without drawing a spacetime diagram, determine whether the line connecting each each pair of events, is timelike, spacelike, or null.*

(b) *Draw these events on a spacetime diagram.*

(c) *Find a reference frame in which two of the events are simultaneous. What is the relative speed between this reference frame and the lab?*

(d) *Determine the coordinates A', B', C' of each event in the new reference frame.*

(e) *Redraw the diagram in this reference frame.*

(a) The (squared) spacetime interval between each pair of points can be calculated algebraically. We have

$$
\begin{aligned}
|AB|^2 &= (4 - 1)^2 - (7 - 2)^2 = 3^2 - 5^2 = -16 < 0, \quad (7.2) \\
|BC|^2 &= (6 - 4)^2 - (5 - 7)^2 = 1^2 - 1^2 = 0, \quad\quad\quad (7.3) \\
|AC|^2 &= (6 - 1)^2 - (5 - 2)^2 = 5^2 - 3^2 = +16 > 0. \quad (7.4)
\end{aligned}
$$

Thus, AB is timelike, BC is null, and AC is spacelike.

(b) A spacetime diagram showing the three events is given in Figure 7.1.

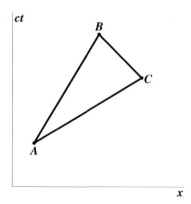

FIGURE 7.1. The spacetime diagram for the three events in the lab frame.

(c) Simultaneous events are spacelike separated, so the only possibility is events A and C. The (hyperbolic) angle ψ between AC and horizontal can be determined from

$$\tanh \psi = \frac{5-2}{6-1} = \frac{3}{5}, \tag{7.5}$$

which implies that the two reference frames have a relative speed of $\frac{3}{5}$ the speed of light.

(d) *Geometric solution:* The trajectory AC can be identified with the hypotenuse of the right triangle in Figure 5.5d. From looking at this 3–4–5 triangle, we see that

$$\cosh \psi = \frac{5}{4}, \qquad \sinh \psi = \frac{3}{4}, \tag{7.6}$$

so the Lorentz transformation *to* the lab frame is

$$\Lambda = \begin{pmatrix} \cosh \psi & \sinh \psi \\ \sinh \psi & \cosh \psi \end{pmatrix} = \begin{pmatrix} \frac{5}{4} & \frac{3}{4} \\ \frac{3}{4} & \frac{5}{4} \end{pmatrix}. \tag{7.7}$$

Since we want the transformation *from* the lab frame, we have:

$$A' = \Lambda^{-1}A = \left(-\tfrac{1}{4}, \tfrac{7}{4}\right), \tag{7.8}$$
$$B' = \Lambda^{-1}B = \left(-\tfrac{1}{4}, \tfrac{23}{4}\right), \tag{7.9}$$
$$C' = \Lambda^{-1}C = \left(\tfrac{15}{4}, \tfrac{7}{4}\right), \tag{7.10}$$

where A, B, C are to be thought of as column vectors when performing the matrix multiplication.

Algebraic solution: We seek a Lorentz transformation of the form

$$\widetilde{\Lambda} = \begin{pmatrix} \cosh\theta & \sinh\theta \\ \sinh\theta & \cosh\theta \end{pmatrix}. \tag{7.11}$$

Acting on events A and C, we obtain

$$A' = \widetilde{\Lambda}A, \qquad C' = \widetilde{\Lambda}C, \tag{7.12}$$

so that

$$A' = (1\cosh\theta + 2\sinh\theta, 1\sinh\theta + 2\cosh\theta), \tag{7.13}$$
$$C' = (6\cosh\theta + 5\sinh\theta, 6\sinh\theta + 5\cosh\theta). \tag{7.14}$$

But if A' and C' are to be simultaneous, we must have

$$\sinh\theta + 2\cosh\theta = 6\sinh\theta + 5\cosh\theta, \tag{7.15}$$

which implies that

$$\tanh\theta = \frac{5-2}{1-6} = -\frac{3}{5}, \tag{7.16}$$

in agreement with the geometric computation above. That is, we have shown that $\widetilde{\Lambda} = \Lambda^{-1}$. The rest of the computation is the same as before.

(e) A spacetime diagram showing the three events in the rotated frame is given in Figure 7.2.

2. *The speed of a typical GPS satellite is* $4000 \; \frac{m}{s}$.

   (a) *Determine this speed as a fraction of the speed of light, that is, compute* $\frac{v}{c}$.

   (b) *Determine the correction factor* $\gamma = \dfrac{1}{\sqrt{1-\frac{v^2}{c^2}}} = \cosh\alpha$.

   (c) *By how much would a clock on this satellite differ from a clock on Earth at the end of 24 hours?*

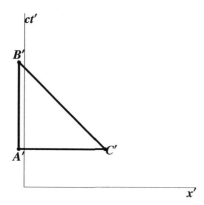

FIGURE 7.2. The spacetime diagram for the three events in the rotated frame.

(d) *Is this difference enough to impact GPS measurements?*

(a) Straightforward division yields $\tanh\alpha = \frac{v}{c} \approx 1.3 \times 10^{-5}$.

(b) We can approximate the square root using the binomial theorem as

$$\gamma = \cosh\alpha = (1-\tanh\alpha)^{-1/2} \approx 1+\frac{1}{2}\tanh^2\beta \approx 1+0.8\times10^{-10}$$
$$(7.17)$$

so the correction is roughly one part in 10^{10}.

(c) Approximating 24 hours as 10^5 seconds, the clocks would differ by roughly 10^{-4} nanoseconds after 24 hours.

(d) Yes, this matters! Light travels roughly 3 kilometers in 10^{-4} nanoseconds! Would you accept that accuracy in your GPS results?

Try the next problem on your own; no solution is provided.

3. *You are driving along the highway at 100 kilometers per hour when you pass your friend Gal's house, who is sitting on the front porch. Thirty seconds later, you see a lightning bolt hit the ground two kilometers ahead of you. At that very instant, Gal's cell phone rings. Some time later, you each hear the thunder clap from the bolt.*

(a) *Draw a spacetime diagram with respect to Gal's frame. Draw and label worldlines for you and Gal. Draw and label events: you passing Gal's house, the lightning strike, the cell phone ring, you seeing light from the lightning, Gal seeing the lightning, you hearing the thunder, Gal hearing the thunder.*

(b) *Determine the spacetime coordinates for each event from Gal's perspective. Label the axes of your spacetime diagram with these values.*

(c) *Draw a new spacetime diagram with respect to your reference frame. Draw and label the same worldlines and events.*

(d) *Determine the spacetime coordinates for each event from your perspective. Label the axes of your spacetime diagram with these values.*

7.2   Practice

4. *In a laboratory experiment a muon is observed to travel 800 meters before disintegrating. The lifetime of a muon is 2×10^{-6} seconds, so the speed must be*

$$v = \frac{800 \text{ m}}{2 \times 10^{-6} \text{ s}} = 4 \times 10^8 \text{ m/s}, \qquad (7.18)$$

which is faster than the speed of light. Identify the error in this computation and find the actual speed of the muon.

The lifetime of the muon must be measured in its own rest frame. Geometrically, we have the triangle in Figure 7.3, whose *hypotenuse* is $2c \times 10^{-6} = 600$ meters. The remaining leg of this 3–4–5 triangle is easily seen to be 1000 meters, from which we can read off

$$\frac{v}{c} = \tanh \alpha = \frac{4}{5}. \qquad (7.19)$$

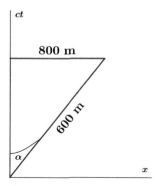

FIGURE 7.3. The spacetime diagram for a traveling muon.

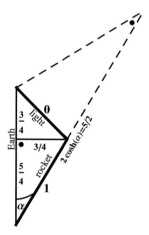

FIGURE 7.4. The spacetime diagram for the rocket ship. The dots indicate right angles.

5. *A rocket ship leaves Earth at a speed of $\frac{3}{5}c$. When a clock on the rocket says 1 hour has elapsed, the rocket sends a light signal back to Earth.*

(a) *According to Earth clocks, when was the signal sent?*

(b) *According to Earth clocks, how long after the rocket left did the signal arrive back on Earth?*

(c) *According to the rocket observer, how long after the rocket left did the signal arrive back on Earth?*

All of these questions can be answered by drawing a spacetime diagram, as shown in Figure 7.4. We again have a 3–4–5 triangle, with hypotenuse of length 1 and $\tanh\alpha = \frac{3}{5}$. We therefore have $\cosh\alpha = \frac{5}{4}$ (the vertical leg) and $\sinh\alpha = \frac{3}{4}$ (the horizontal leg).

(a) The signal was sent after $\frac{5}{4}$ hours according to Earth clocks.

(b) The signal was received $\frac{3}{4}$ hours after it was sent, or two hours after the rocket left, again according to Earth clocks.

(c) We consider the large right triangle in Figure 7.4, also with hyperbolic angle α, and with vertical side 2. This side is the hypotenuse, since the right angle is in the upper right corner. We are trying to find the remaining side, which is $2\cosh\alpha = \frac{5}{2}$, so the signal reaches Earth $\frac{5}{2}$ hours after the rocket left, according to its own clocks.

6. *A Lincoln Continental is twice as long as a VW Beetle when they are at rest. As the Lincoln overtakes the VW, going through a speed trap, a stationary policeman observes that they both have the same length. The VW is going at half the speed of light. How fast is the Lincoln going?*

Suppose the speed (really the rapidity) of the VW is given by $\tanh\psi$ and that of the Lincoln by $\tanh\alpha$. A spacetime diagram showing the worldlines of the front and back of each car is shown in Figure 7.5(a). Since the cars appear to be the same length, these

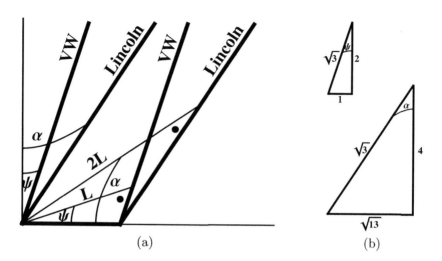

(a)   (b)

FIGURE 7.5. (a) The spacetime diagram for the Lincoln and the VW (not to scale). (b) Two helpful hyperbolic right triangles.

worldlines intersect along the horizontal axis. The "true" lengths of the cars, namely L and $2L$, respectively, are also shown. Comparing the right triangles shown, whose right angles are indicated with dots and which share a common horizontal leg, we see that

$$\frac{L}{\cosh\psi} = \frac{2L}{\cosh\alpha}, \tag{7.20}$$

so that

$$\cosh\alpha = 2\cosh\psi. \tag{7.21}$$

But we are given that $\tanh\psi = \frac{1}{2}$, so that, by using the triangles shown in Figure 7.5(b), we have in turn that

$$\cosh\psi = \frac{2}{\sqrt{3}}, \tag{7.22}$$

$$\cosh\alpha = \frac{4}{\sqrt{3}}, \tag{7.23}$$

$$\tanh\alpha = \frac{\sqrt{13}}{4}, \tag{7.24}$$

so that the Lincoln is moving at $\frac{\sqrt{13}}{4}$ the speed of light.

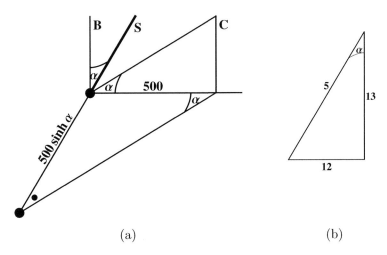

(a)   (b)

FIGURE 7.6. (a) Spacetime diagram for the scientist's analysis (not to scale), showing the worldlines of the brother (B), scientist (S), and clairvoyante (C). (b) Helpful right triangle.

7. *Sophie Zabar, clairvoyante, cried out in pain at precisely the instant her twin brother, 500 km away, hit his thumb with a hammer. A skeptical scientist observed both events from an airplane traveling at $\frac{12}{13}c$ to the right. Which event occurred first, according to the scientist? How much earlier was it?*

This situation is shown in Figure 7.6, from the shared reference frame of the clairvoyante (C) and her brother (B). The worldline of the scientist (S) is tilted to the right with hyperbolic angle α; his lines of simultaneity make the same angle with the horizontal axis. His time is measured along his own worldline, so we are looking for the "distance" between the two heavy dots. Triangle trigonometry does the rest; this leg has length in kilometers given by

$$500 \sinh \alpha = 1200, \tag{7.25}$$

where we have made use of the smaller 5–12–13 triangle shown in the figure to compute $\sinh \alpha = \frac{12}{5}$ from $\tanh \alpha = \frac{12}{13}$. Converting this distance to a time by dividing by c, we conclude that the scientist observes the brother hit his thumb 4×10^{-3} seconds *before* Sophie cried out.

7.3   The Getaway

8.   *The outlaws are escaping in their getaway car, which moves at $\frac{3}{4}c$, chased by the police, moving at only $\frac{1}{2}c$. Realizing they can't catch up, the police attempt to shoot out the tires of the getaway car. Their guns have a muzzle velocity (speed of the bullets relative to the gun) of $\frac{1}{3}c$.*

(a) *Does the bullet reach its target according to Galileo?*

(b) *Does the bullet reach its target according to Einstein?*

(c) *Verify that your answer to part (b) is the same in all four reference frames: ground, police, outlaws, and bullet.*

(a) In Newtonian physics, we add the speed of the bullet $\left(\frac{1}{3}c\right)$ to that of the police $\left(\frac{1}{2}c\right)$ to compute the speed of the bullet with respect to the ground, obtaining $\frac{5}{6}c$, which is greater than the speed of the outlaws $\left(\frac{3}{4}c\right)$. Yes, the bullet reaches its target according to Galileo.

(b) Denote the rapidities of the police and outlaws with respect to the ground by ψ and α, respectively, and let the rapidity of the bullet with respect to the police be η. Then we have

$$\tanh \psi = \frac{1}{2}, \tag{7.26}$$

$$\tanh \alpha = \frac{3}{4}, \tag{7.27}$$

$$\tanh \eta = \frac{1}{3}. \tag{7.28}$$

In special relativity, we add the rapidities (hyperbolic angles), so that the speed of the bullet with respect to the

	Ground	Police	Outlaws	Bullet
Ground	0	$\frac{1}{2}$	$\frac{3}{4}$	$\frac{5}{7}$
Police	$-\frac{1}{2}$	0	$\frac{2}{5}$	$\frac{1}{3}$
Outlaws	$-\frac{3}{4}$	$-\frac{2}{5}$	0	$-\frac{1}{13}$
Bullet	$-\frac{5}{7}$	$-\frac{1}{3}$	$\frac{1}{13}$	0

TABLE 7.1. The relative speeds of the ground, police, outlaws, and bullet with respect to each other.

ground is given by

$$\frac{v_b}{c} = \tanh(\psi + \eta) = \frac{\tanh\psi + \tanh\eta}{1 + \tanh\psi \tanh\eta} = \frac{5}{7}. \quad (7.29)$$

Thus, v_b is less than the speed of the outlaws $(\frac{3}{4}c)$, and the bullet does not reach its target, according to Einstein.

(c) By performing a similar computation in each case, we obtain Table 7.1, in which all speeds are given as fractions of c.

7.4 Angles Are Not Invariant

9. *The mast of a sailboat leans at an angle θ (measured from the deck) toward the rear of the boat.*

   (a) *An observer on the dock sees the boat go by at speed $v \ll c$ (so you do not need to use relativity to do this problem). What angle does the observer say the mast makes?*

   (b) *A child on the boat throws a ball into the air at the same angle θ. What angle does the observer on the dock say the ball makes with the deck? (Ignore the*

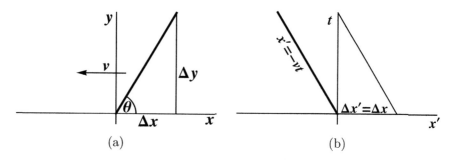

(a)   (b)

FIGURE 7.7. (a) The angle made by the mast of a moving sailboat.
(b) The speed of the mast. (Primed coordinates refer to the dock.)

subsequent influence of gravity on the ball—this ques-
tion is only about the initial angle when the ball leaves
the child's hand.)

(c) *A spaceship goes past the dock at speed v. An antenna*
is mounted on its hull at an angle θ with the (hori-
zontal) hull. What angle does the observer on the dock
say the antenna makes?

(d) *A spotlight is mounted on the spaceship so that its*
beam makes an angle θ with the hull. What angle does
the observer on the dock say the beam makes with the
hull?

(a) See Figure 7.7. The angles are the same, since both observers agree on the lengths of the sides of the right
triangle made by the mast and the deck. Using primed
coordinates to refer to the dock, and unprimed coordinates for the sailboat, we have

$$\Delta x' = \Delta x, \tag{7.30}$$
$$\Delta y' = \Delta y, \tag{7.31}$$

so that

$$\tan \theta' = \frac{\Delta y'}{\Delta x'} = \frac{\Delta y}{\Delta x} = \tan \theta. \tag{7.32}$$

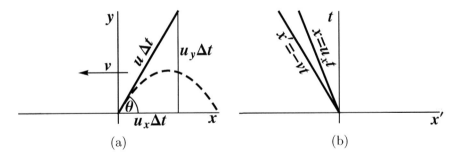

FIGURE 7.8. (a) The angle made by a ball thrown from a moving sailboat. (b) The speed of the ball.

(b) See Figure 7.8. The right triangle used to determine this angle now involves velocities, not lengths, and velocities are different as measured by the two observers. We now have

$$u'_x = u_x - v, \tag{7.33}$$
$$u'_y = u_y, \tag{7.34}$$

so that

$$\tan \theta' = \frac{u'_y}{u'_x} = \frac{u_y}{u_x - v} = \frac{\tan \theta}{1 - \frac{v}{u\cos\theta}} = \frac{u\sin\theta}{u\cos\theta - v}. \tag{7.35}$$

(c) See Figure 7.9. Now we have to take length contraction into account. Thus,

$$\Delta x' = \frac{\Delta x}{\cosh \alpha}, \tag{7.36}$$
$$\Delta y' = \Delta y, \tag{7.37}$$

so that

$$\tan \theta' = \frac{\Delta y'}{\Delta x'} = \frac{\Delta y}{\Delta x} \cosh \alpha = \tan \theta \cosh \alpha > \tan \theta. \tag{7.38}$$

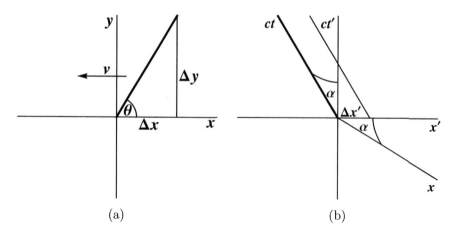

FIGURE 7.9. (a) The angle made by an antenna on a moving spaceship. (b) The Lorentz transformation between frames.

(d)   See Figure 7.10. We must now deal with both the hyperbolic angle due to the spaceship's velocity and the Euclidean angle formed by the beam of light, which does *not* travel at the speed of light *in the direction of the spaceship*. We have, as usual, $\frac{v}{c} = \tanh\alpha$, where v is the speed of the spaceship. The component of the velocity of the beam of light *in this direction* is given by

$$\tanh\psi = \cos\theta, \qquad\qquad (7.39)$$

as can be seen by using Figure 7.10(a). By using Figure 7.10(b), we see that

$$-\cos\theta' = \tanh(\alpha-\psi) = \frac{\tanh\alpha - \tanh\psi}{1 - \tanh\alpha\tanh\psi} = \frac{v - c\cos\theta}{c - v\cos\theta}.$$
$$(7.40)$$

7.5   Interstellar Travel

10.   *Alpha Centauri is roughly 4 lightyears from Earth. Dr. X travels (at constant velocity) from Earth to Al-*

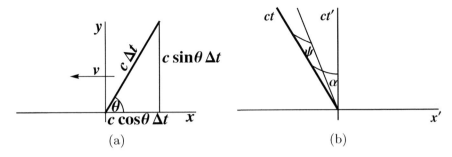

FIGURE 7.10. (a) The (Euclidean) angle made by the beam of a moving searchlight. (b) The (hyperbolic) angle between the two reference frames.

pha Centauri in 3 years. Immediately upon her arrival at Alpha Centauri, she turns on a powerful laser aimed at Earth. (Ignore the Earth's motion.)

(a) *How fast did Dr. X travel?*

(b) *How far did Dr. X travel? (That is, how far does she think Alpha Centauri is from Earth?)*

(c) *How long after Dr. X's departure do observers on Earth find out she arrived safely?*

(d) *Draw a spacetime diagram showing the worldlines of Dr. X, Earth, and the laser beam. Clearly indicate which events correspond to departure, arrival/turning on the laser, and the receipt of the laser signal on Earth.*

See the spacetime diagram in Figure 7.11, which answers part (d).

(a) The known data are that Alpha Centauri is 4 lightyears from Earth, and that Dr. X travels for 3 years *as measured by her own clock*. The lower triangle in Figure 7.11 therefore has horizontal leg $\Delta x = 4$ lightyears and hypotenuse $c\Delta\tau = 3$ lightyears. The remaining (vertical) leg of this triangle therefore has length 5 lightyears, from which we can read off

$$\frac{v}{c} = \tanh\alpha = \frac{4}{5}. \tag{7.41}$$

FIGURE 7.11. The spacetime diagram for Dr. X.

(b) This same triangle gives us a Lorentz contraction factor
     of
     $$\cosh \alpha = \frac{5}{3}, \qquad (7.42)$$
     from which it follows that Dr. X thinks that the distance
     from Earth to Alpha Centauri is
     $$\ell' = \frac{\ell}{\cosh \alpha} = \frac{4}{\frac{5}{3}} = \frac{12}{5} \qquad (7.43)$$
     Alternatively, if Dr. X travels for 3 years at $\frac{4}{5}$ the speed
     of light, she must have traveled $\frac{12}{5}$ lightyears.

(c) We already know that the vertical leg of the lower triangle
     has length 5, indicating that observers on Earth believe
     Dr. X arrives 5 years after departure. The laser signal
     will take 4 years to travel 4 lightyears, for a total time
     difference of 9 years.

7.6   Observation

11. *Two observers are moving at relative speed $\frac{v}{c} = \frac{4}{5}$. Nine
     minutes after they meet at noon (by their own clock), they
     use a telescope to look at each other's clock. What time
     do they see displayed on the other clock?*

It is essential to realize that this problem is about *seeing*, not *observing*. "Seeing" involves the use of light rays to carry information, whereas "observing" assumes a family of observers scattered throughout spacetime who can take notes and compare their observations afterward. See Section 2.1.

But this is the same problem as in Section 7.5 in disguise! See the spacetime diagram in Figure 7.11, converting lightyears to light minutes, and ignoring the labels. In the reference frame shown, the observer at rest takes nine minutes to travel along the trajectory labeled "Earth". When she looks in her telescope at that moment, she sees the light arriving along the trajectory labeled "laser," and therefore "sees" the other observer's clock at the time the light was emitted. The other observer ("Dr. X") has traveled three minutes to reach that point. This argument is the same for both observers, so each sees the other clock display 12:03 PM.

7.7   Cosmic Rays

12. *Consider muons produced by the collision of cosmic rays with gas nuclei in the atmosphere 60 kilometers above the surface of the earth, which then move vertically downward at nearly the speed of light. The half-life before muons decay into other particles is 1.5 microseconds $(1.5 \times 10^{-6}$ s).*

    (a) *Assuming it doesn't decay, how long would it take a muon to reach the surface of the earth?*

    (b) *Assuming there were no time dilation, approximately what fraction of the muons would reach the earth without decaying?*

    (c) *In actual fact, roughly $\frac{1}{8}$ of the muons would reach the earth. How fast are they going?*

    (a) Without much loss of accuracy, assume the muons travel at the speed of light. Then it takes them

$$\frac{60 \text{ km}}{3 \times 10^8 \text{ m/s}} = 200 \ \mu s \qquad (7.44)$$

    to reach the earth.

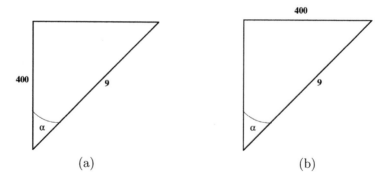

FIGURE 7.12. Hyperbolic triangles for the cosmic ray example. (a) Diagram for first solution. (b) Diagram for second solution.

(b) Since 200 μs is $\dfrac{200}{1.5} = \dfrac{400}{3}$ half-lives, only $2^{-\frac{400}{3}}$ of the muons reach the earth.

(c) *First solution:* The given fraction corresponds to 3 half-lives, since $\frac{1}{8} = 2^{-3}$. Thus, the time is dilated by a factor of $\frac{400/3}{3}$, so that

$$\cosh \alpha = \frac{400}{9}. \tag{7.45}$$

But

$$\frac{v}{c} = \tanh \alpha = \frac{\sqrt{400^2 - 9^2}}{400} \approx .99974684. \tag{7.46}$$

See Figure 7.12(a).

Second solution: The first solution argument assumes $v \approx c$. (This was used to obtain the figure 200 μs in part (a), which is hence only an approximation.)

A more accurate argument would use the fact that the muons travel 60 km in $3 \times 1.5 \times 10^{-6}$ s (3 half-lives) *of proper time.* Thus,

$$\sinh \alpha = \frac{(60 \text{ km})(1000 \text{ m/km})}{(4.5 \times 10^{-6} \text{ s})(3 \times 10^8 \text{ m/s})} = \frac{400}{9}, \qquad (7.47)$$

so that

$$\frac{v}{c} = \tanh \alpha = \frac{400}{\sqrt{400^2 + 9^2}} \approx .99974697. \qquad (7.48)$$

See Figure 7.12(b).

It is important to realize not only that the second solution is more accurate than the first (assuming sufficient accuracy in the original data), but also that that the "shortcut" used in the first solution is justified.[1]

7.8 Doppler Effect

13. *A rocket sends out flashes of light every two seconds in its own rest frame, which you receive every 4 seconds. How fast is the rocket going?*

First solution: This situation is shown in Figure 7.13(a). To find the hyperbolic angle α, draw a horizontal line as shown in the enlargement in Figure 7.13(b), resulting in the system of equations[2]

$$\tanh \alpha = \frac{x}{4 - x}, \qquad (7.50)$$

$$(4 - x)^2 - x^2 = 2^2, \qquad (7.51)$$

which is easily solved for $x = \frac{3}{2}$, so that $\frac{v}{c} = \tanh \alpha = \frac{3}{5}$.

---

[1] This can be made rigorous by using a power series expansion. Equivalently, by simply redoing the computation using the approximation obtained in part (c) to recalculate part (a), the correct answer can be obtained to any desired accuracy. The speed at which this iterative procedure converges to the exact answer justifies having made the approximation in the first place.

[2] This method can be be used to derive the Doppler shift formula in general, yielding

$$\frac{v}{c} = \frac{\lambda^2 - \lambda'^2}{\lambda^2 + \lambda'^2}, \qquad (7.49)$$

which is equivalent to Equation (6.17); in this example, $\lambda = 4$ and $\lambda' = 2$.

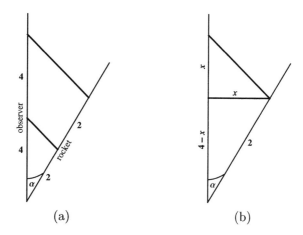

FIGURE 7.13. Computing Doppler shift. (a) Two flashes of light. (b) An enlarged view showing just one flash of light.

Second solution: Insert $\lambda = 4$ and $\lambda' = 2$ into Equation (6.17) and solve for $\frac{v}{c}$.

8

Paradoxes

In which impossible things are shown to be possible.

8.1   Special Relativity Paradoxes

It is easy to create seemingly impossible scenarios in special relativity by playing on the counterintuitive nature of observer-dependent time. These scenarios are usually called paradoxes because they seem to be impossible. Yet there is nothing paradoxical about them.

The best way to resolve these paradoxes is to draw a good spacetime diagram, which in turn requires careful reading of the problem, always making sure to associate the given information with a particular reference frame. A single spacetime diagram suffices to determine what *all* observers see. It is nevertheless instructive to draw separate spacetime diagrams for each observer, making sure that they all agree.

In this chapter, we discuss the two most famous paradoxes in special relativity, the pole and barn paradox and the twin paradox. We also briefly discuss the more subtle aspects of flying manhole covers.

8.2   The Pole and Barn Paradox

A 20-foot pole is moving toward a 10-foot barn fast enough that the pole appears to be only 10 feet long. As soon as both ends of

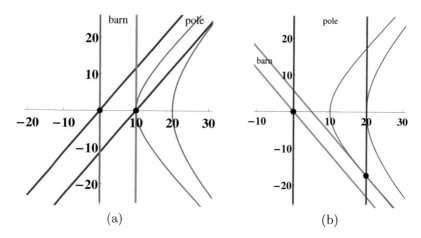

FIGURE 8.1. The pole and barn paradox, (a) in the barn's reference frame and (b) in the pole's reference frame. In each diagram, the heavy straight lines represent the ends of the pole and the lighter straight lines represent the front and back of the barn. The hyperbolas of "radius" 10 and 20 are also shown. The dot at the origin labels the event where the *back* of the pole *enters* the barn, and the other dot labels the event where the *front* of the pole *leaves* the barn.

the pole are in the barn, the doors are slammed shut. How can a 20-foot pole fit into a 10-foot barn?

This is the beginning of the pole and barn paradox. It's bad enough to try to imagine what happens to the pole when it suddenly stops and finds itself in a barn that is too small. But what does the pole see?

We know that length contraction is symmetric, so if the barn sees the pole shortened by a factor of 2, then the pole sees the barn shortened by the same factor of 2. This factor is just $\cosh \alpha$, where $|\alpha|$ is the same in both cases. So the 20-foot pole sees this 5-foot barn approaching. No way is the pole going to fit in the barn!

The spacetime diagrams for this situation are shown in Figures 8.1(a) and (b), as seen in the barn's reference frame and in the pole's reference frame, respectively. The two dots represent the events of closing the barn doors when the ends of the pole are even with the corresponding door.

In the *barn's* frame (Figure 8.1(a)), these events happen simultaneously, so it is possible in principle to trap the pole in the barn by shutting the doors "at the same time." (We omit speculation about what happens when the pole hits the closed door!)

In the *pole's* frame (Figure 8.1(b)), the exit door is closed long before the rear of the pole enters the barn. Assuming the pole keeps going, for instance by virtue of the door opening again, then the entrance door is closed much later, when the rear of the pole finally gets there. The pole "thinks" it is silly to try to catch it by waiting to close the entrance door until most of the pole has already escaped through the exit door!

8.3 The Twin Paradox

One twin travels 24 lightyears to star X at speed $\frac{24}{25}c$; her twin brother stays home on Earth. When the traveling twin gets to star X, she immediately turns around, and returns at the same speed. How long does each twin think the trip took?

Star X is 24 lightyears away, so, according to the twin at home, it takes her 25 years to get there, and 25 more to return, for a total of 50 years away from Earth. But the traveling twin's clock runs slow by a factor of

$$\cosh \alpha = \sqrt{\frac{1}{1 - \tanh^2 \alpha}} = \frac{25}{7}.$$

Thus, according to the traveling twin, it takes her only 7 years each way, so that she has aged only 14 years while her brother has aged 50.

This result is, in fact, correct, and represents a sort of time travel into the future: it takes the traveling twin 14 years to get 50 years into Earth's future. (Unfortunately, there's no way to get back.)

But wait a minute. The traveling twin should see her brother's clock run slow by the same factor of $\frac{25}{7}$. So when 7 years of her time elapse, she thinks her brother has aged only $\frac{49}{25} \approx 2$ years. Her

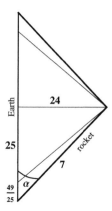

FIGURE 8.2. Spacetime diagram for the twin paradox (not to scale). The vertical line represents the world line of the stay-at-home twin, and the heavy diagonal lines show the world line of the traveling twin.

brother should therefore have aged only 4 years when she returns, from her perspective.

That cannot be right. Either her brother is 4 years older or he is 50 years older. Both siblings must surely agree on that!

The easiest way to resolve this paradox is to draw a single spacetime diagram showing the entire trip in the reference frame of the stay-at-home twin, as shown in Figure 8.2. The lower half of the figure is a hyperbolic triangle with $\tanh \alpha = \frac{24}{25}$, and hypotenuse 7 years. The remaining diagonal lines are lines of constant time for the traveling twin, at the point of turnaround, while going and returning, respectively. There is another right triangle containing the hyperbolic angle α; the right angle is at the point of turnaround, and the hypotenuse is $\frac{7}{\cosh \alpha} = \frac{49}{25}$ years, the age of the stay-at-home twin "when" the traveling twin turns around.

How much does each twin age? Simply measure the length of their world lines. Intervals are invariant, so it doesn't matter how you compute them. The clear answer is that the brother has indeed aged 50 years, while the sister has aged only 14 years.

So what was wrong with the argument that the brother should have aged only four years? There are really three reference frames here: Earth, going, and returning. The "going" and "returning" frames yield different times on Earth for the turnaround—and these times differ by precisely $\frac{1152}{25} \approx 46$ years. From this point of view, it takes the traveling twin 46 years to turn around.

One difference between the twins is that the traveling twin is *not* in an inertial frame—she is in two inertial frames, but must accelerate in order to switch from one to the other. This breaks the symmetry between the two twins.

However, this is not really the best way to explain this paradox. It is possible to remove this particular asymmetry by assuming the universe is closed, so that the traveling twin doesn't need to turn around. A simplified version of this is to put the problem on a cylinder [Dray 90]. It turns out that this introduces another sort of asymmetry, but there is a simpler way to look at it.

The amount an observer ages is just the timelike interval measured along his or her worldline. We used this argument here. This approach also works for curved worldlines, corresponding to noninertial observers—except that one must integrate the infinitesimal timelike interval $d\tau = \sqrt{dt^2 - ds^2}$ along the worldline.

A little thought leads to the following remarkable result: *The timelike line connecting two events (assuming there is one) is the longest path from one to the other.* (Think about it. Use this line as the t axis. Then any other path has a nonzero contribution from the change in x—which *decreases* the hyperbolic length of the path, and hence the time taken.)

8.4 Manhole Covers

There are two well-known paradoxes involving manhole covers that illustrate some unexpected implications of special relativity. In the first, a 2-foot manhole cover approaches a 2-foot manhole at relativistic velocity. Since the hole sees the cover as much smaller than two feet long, the cover must fall into the manhole. It does.

But what does the cover see? It sees a very small hole rushing at it. No way is this enormous manhole cover going to fit into this small hole.

The resolution of this paradox requires careful consideration of what it means for something to begin to fall, and is left to the reader.

There is also a higher-dimensional version of this problem, without the complication of falling—that is, without gravity. Suppose the manhole cover is moving horizontally as before, but now the hole is in a metal sheet that is rising up to meet it. Again, from the point of view of the hole, the cover is very small and so—if the timing is right—the cover will pass through the hole. It does.

But what does the cover see? It again sees a very small hole rushing at it. How do you get a big object through a small hole? This time we must consider what it means for the cover to "pass through" the hole. A further analysis of this paradox, using three-dimensional spacetime diagrams, is given in Section 14.2.

Anyone who successfully resolves these two paradoxes will realize that some properties of materials that we take for granted are quite impossible in special relativity.

9

Relativistic Mechanics

In which it is shown that mass is energy.

9.1 Proper Time

Let τ denote time as measured by a clock carried by an observer moving at constant speed $u = \tanh\psi$ with respect to the given frame. We call τ "wristwatch time" [Taylor and Wheeler 92], or, more formally, *proper time*. Such a clock simply measures the "length" of the observer's worldline. But this length can be measured in any reference frame.

The (infinitesimal) geometry of this situation is shown in Figure 9.1. Since $d\tau$ is the length of the hypotenuse, it follows immediately that

$$d\tau = \frac{1}{\cosh\psi}\,dt. \tag{9.1}$$

Equivalently, since the position of the moving observer doesn't change in the moving observer's own rest frame, we have, from the invariance of the interval, that

$$dx^2 - c^2 dt^2 = 0 - c^2 d\tau^2, \tag{9.2}$$

so that

$$d\tau^2 = \left(1 - \frac{1}{c^2}\left(\frac{dx}{dt}\right)^2\right)dt^2, \tag{9.3}$$

or, equivalently,

$$d\tau = \sqrt{1 - \frac{u^2}{c^2}}\,dt = \frac{1}{\gamma}\,dt = \frac{1}{\cosh\psi}\,dt. \tag{9.4}$$

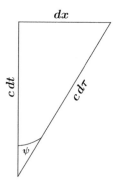

FIGURE 9.1. The geometry of proper time.

This computation yields the same answer (for τ, not for t) in any reference frame; proper time is independent of reference frame.

9.2   Velocity

Consider the *ordinary velocity* of a moving object, defined by

$$u = \frac{d}{dt} x. \tag{9.5}$$

This quantity transforms in a complicated way, since

$$\frac{1}{c}\frac{dx'}{dt'} = \frac{\frac{1}{c}\frac{dx}{dt} - \frac{v}{c}}{1 - \frac{v}{c^2}\frac{dx}{dt}}, \tag{9.6}$$

both the numerator and the denominator need to be transformed. Equation (9.6) is just the Einstein addition formula for velocities, which we have therefore independently derived by using Lorentz transformations. But that's not quite what we are looking for here.

The invariance of proper time suggests that we should instead differentiate with respect to proper time, since of course

$$\frac{d}{d\tau} x' = \frac{dx'}{d\tau}. \tag{9.7}$$

In other words, the operator $\frac{d}{d\tau}$ pulls through the Lorentz transformation; only the numerator is transformed when changing reference frames.

Furthermore, the same argument can be applied to t, which suggests that there are (in two dimensions) two components to the velocity. We therefore consider the "2-velocity"

$$\boldsymbol{u} = \frac{d}{d\tau}\begin{pmatrix} ct \\ x \end{pmatrix} = \begin{pmatrix} c\frac{dt}{d\tau} \\ \frac{dx}{d\tau} \end{pmatrix}. \tag{9.8}$$

But since

$$dt = \cosh\psi\, d\tau \tag{9.9}$$

and

$$dx^2 - c^2 dt^2 = -c^2 d\tau^2, \tag{9.10}$$

we also have

$$dx = c\sinh\psi\, d\tau, \tag{9.11}$$

so that

$$\boldsymbol{u} = c\begin{pmatrix} \cosh\psi \\ \sinh\psi \end{pmatrix}. \tag{9.12}$$

Note that $\frac{1}{c}\boldsymbol{u}$ is a *unit* vector; that is,

$$\frac{1}{c^2}\boldsymbol{u}\cdot\boldsymbol{u} = -1. \tag{9.13}$$

Further,

$$\frac{u}{c} = \frac{dx}{c\, dt} = \tanh\psi, \tag{9.14}$$

as expected.

9.3 Conservation Laws

Suppose that Newtonian momentum is conserved in a given frame; that is,

$$\sum m_i v_i = \sum M_j V_j. \tag{9.15}$$

(Both of these sums would be zero in the center-of-mass frame.)
Changing to another frame moving with respect to the first at
speed v, we have

$$v_i = v'_i + v, \tag{9.16}$$
$$V_j = V'_j + v, \tag{9.17}$$

so that

$$\sum m_i(v'_i + v) = \sum M_j(V'_j + v). \tag{9.18}$$

We therefore see that

$$\sum m_i v'_i = \sum M_j V'_j \iff \sum m_i = \sum M_j. \tag{9.19}$$

That is, momentum is conserved in *all* inertial frames provided it
is conserved in one frame *and* mass is conserved.

Repeating the computation for kinetic energy, we obtain, starting from

$$\frac{1}{2} \sum m_i v_i^2 = \frac{1}{2} \sum M_j V_j^2, \tag{9.20}$$

that

$$\frac{1}{2} \sum m_i(v'_i + v)^2 = \frac{1}{2} \sum M_j(V'_j + v)^2. \tag{9.21}$$

Expanding out these expressions, we discover that (kinetic) energy
is conserved in all frames provided it is conserved in one frame *and*
both mass and momentum are conserved.

The situation in special relativity is quite different.

Consider first the momentum defined by the ordinary velocity,
namely,

$$p = mu = m\frac{dx}{dt}. \tag{9.22}$$

This momentum is *not* conserved.

We use instead the momentum defined by the 2-velocity, which
is given by

$$p = m\frac{dx}{d\tau} = mc\sinh\psi. \tag{9.23}$$

Suppose now that, as seen in a particular inertial frame, the total momentum of a collection of particles is the same before and after some interaction; that is,

$$\sum m_i c \sinh \psi_i = \sum M_j c \sinh \Psi_j. \tag{9.24}$$

Consider now the same situation as seen by another inertial reference frame, moving with respect to the first with speed

$$v = c \tanh \alpha. \tag{9.25}$$

We therefore have

$$\psi_i = \psi_i' + \alpha, \tag{9.26}$$
$$\Psi_j = \Psi_j' + \alpha. \tag{9.27}$$

Inserting Equation (9.26) into the conservation rule (9.24) leads to

$$\sum m_i c \sinh \psi_i' = \sum m_i c \sinh(\psi_i - \alpha)$$
$$= \left(\sum m_i c \sinh \psi_i \right) \cosh \alpha - \left(\sum m_i c \cosh \psi_i \right) \sinh \alpha \tag{9.28}$$

and similarly inserting Equation (9.27) into (9.24) leads to

$$\sum M_j c \sinh \Psi_j' = \left(\sum M_j c \sinh \Psi_j \right) \cosh \alpha - \left(\sum M_j c \cosh \Psi_j \right) \sinh \alpha \tag{9.29}$$

The coefficients of $\cosh \alpha$ in these two expressions are equal due to the assumed conservation of momentum in the original frame. We therefore see that conservation of momentum will hold in the new frame if and only if, in addition, the coefficients of $\sinh \alpha$ agree, namely,

$$\sum m_i c \cosh \psi_i = \sum M_j c \cosh \Psi_j. \tag{9.30}$$

But what is this?

9.4   Energy

This mystery is resolved by recalling that momentum is mass times velocity, and that there is also a "*t*-component" to the velocity. In analogy with 2-velocity, we therefore define "2-momentum" to be

$$\boldsymbol{p} = m \begin{pmatrix} c\frac{dt}{d\tau} \\ \frac{dx}{d\tau} \end{pmatrix} = mc \begin{pmatrix} \cosh\psi \\ \sinh\psi \end{pmatrix}. \tag{9.31}$$

The second term is clearly the momentum, which we denote by p, but what is the first term? If the object is at rest, $\psi = 0$, and the first term is therefore just mc. But Einstein's famous equation,

$$E = mc^2, \tag{9.32}$$

leads us to suspect that this is some sort of energy. In fact, mc^2 is called the *rest energy* or *rest mass*.

   In general, we *define* the energy of an object moving at speed $u = c\tanh\psi$ to be the first component of $\boldsymbol{p}$; that is, we define

$$E := mc^2 \cosh\psi, \tag{9.33}$$
$$p := mc \sinh\psi, \tag{9.34}$$

or equivalently,

$$\boldsymbol{p} = \begin{pmatrix} \frac{1}{c}E \\ p \end{pmatrix}. \tag{9.35}$$

Is this definition reasonable? Consider the case $\frac{u}{c} \ll 1$. Then

$$E = mc^2 \cosh\psi = mc^2\gamma$$
$$= \frac{mc^2}{\sqrt{1 - \frac{u^2}{c^2}}} \tag{9.36}$$
$$\approx mc^2 + \frac{1}{2}mu^2 + \frac{3}{8}m\frac{u^4}{c^2} + \dots .$$

The first term is the rest energy, the next term is the Newtonian kinetic energy, and the remaining terms are relativistic corrections to the kinetic energy.

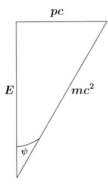

FIGURE 9.2. The geometry of 2-momentum.

The moral is that conservation of 2-momentum is equivalent to both conservation of momentum and conservation of energy, but there is no requirement that the total mass be conserved.

Taking the (squared) norm of the 2-momentum, we obtain

$$-c^2 \boldsymbol{p} \cdot \boldsymbol{p} = E^2 - p^2 c^2 = m^2 c^4. \tag{9.37}$$

The geometry of this situation is shown in Figure 9.2.

Equation (9.37) continues to make sense if $m = 0$, even though the expressions for E and p separately in terms of ψ or γ do not. In fact, γ must approach ∞, or equivalently, $\frac{u^2}{c^2} = 1$, so that $|u| = c$; such particles *always* move at the speed of light.

We therefore conclude that there can be massless particles, that move at the speed of light and satisfy $m = 0$ and

$$E = |p|c \neq 0. \tag{9.38}$$

Photons are examples of such particles; quantum mechanically, one has $E = \hbar \nu$, where ν is the frequency of the light and $\hbar = \frac{h}{2\pi}$ where h is Planck's constant.

9.5 Useful Formulas

The key formulas for analyzing the collision of relativistic particles can all be derived from definitions (9.33) and (9.34).

Taking the difference of squares leads to the key formula (9.37) relating energy, momentum, and (rest) mass, which holds also for massless particles. Rewriting definitions (9.33) and (9.34) leads directly to

$$\gamma = \cosh \psi = \frac{E}{mc^2} = \frac{1}{\sqrt{1 - \frac{u^2}{c^2}}} \tag{9.39}$$

and

$$\sinh \psi = \frac{p}{mc} = \frac{u}{c}\gamma, \tag{9.40}$$

and dividing formula (9.40) by (9.39) yields

$$\tanh \psi = \frac{pc}{E} = \frac{u}{c}. \tag{9.41}$$

9.6   Higher Dimensions

The world we live in has three spatial dimensions, not just one. Special relativity is therefore really a theory in four dimensions, with time being the fourth dimension. The "2-vector" formalism introduced in this chapter naturally generalizes to four dimensions, as discussed further in Section 11.3. Problems in kinematics are then usually expressed in terms of *4-vectors* such as the *4-velocity*, generalizing the 2-velocity defined in Section 9.2, or, more commonly, in terms of the *4-momentum*. Conservation of 4-momentum then replaces both conservation of energy and of (ordinary, 3-dimensional) momentum.

As in Euclidean space, spacetime vectors can be manipulated both geometrically and algebraically. The components of the momentum 2-vector are given algebraically in (9.31), and its geometric interpretation can be inferred from Figure 9.2: It is a (future-pointing!) vector of magnitude mc^2, whose direction is given by the hypotenuse of the triangle shown in the figure.

10

Problems II

10.1 Mass Isn't Conserved

1. *Two identical lumps of clay of (rest) mass m collide head on, with each moving at $\frac{3}{5}c$. What is the mass of the resulting lump of clay?*

First solution: We assume this is an inelastic collision, that is, we do not worry about the details of the actual collision. Conservation of momentum doesn't help here—there is no momentum either before or after the collision. So we need to use conservation of energy. After the collision, there is no kinetic energy, so we have

$$E' = Mc^2. \tag{10.1}$$

Before the collision, we know that the energy of each lump is

$$E = mc^2 \cosh \psi, \tag{10.2}$$

but how do we find ψ? We are given that each lump is moving at $\frac{3}{5}c$, so this means we know

$$\tanh \psi = \frac{3}{5}. \tag{10.3}$$

Yes, we could now use the formula $\cosh \psi = \frac{1}{\sqrt{1-\tanh^2 \psi}}$, but it is easier to use a triangle. Since $\tanh \psi = \frac{3}{5}$, we can scale things so that the legs have "lengths" 3 and 5. Using the hyperbolic Pythagorean theorem, the hypotenuse has length $\sqrt{5^2 - 3^2} = 4$. This is just the triangle in Figure 4.4. Thus,

$$\cosh \psi = \frac{5}{4} \tag{10.4}$$

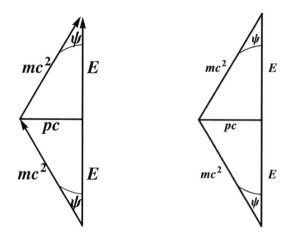

FIGURE 10.1. The momentum diagram for two colliding lumps of clay, first with momentum vectors shown as arrows, then with the arrows suppressed.

so that

$$Mc^2 = E' = 2E = 2mc^2 \cosh \psi = \frac{5}{2} mc^2, \qquad (10.5)$$

and therefore

$$M = \frac{5}{2} m. \qquad (10.6)$$

Second solution: Use conservation of 2-momentum, a vector quantity. That is, draw a *momentum diagram*, similar to a spacetime diagram but with units of either momentum (or, more commonly, energy), showing the 2-momenta both before and after the collision. Figure 10.1 shows such a diagram for the two lumps of clay. The initial momenta, with opposite spatial orientations, are represented by the two diagonal arrows, arranged sequentially; the final momentum is the vector sum of the initial momenta, represented by the vertical arrow. Since all momenta are future-pointing, time-like vectors, such diagrams are often drawn without the arrows, as shown in the second diagram. Simple triangle trigonometry leads to the same computation as in the first solution. Thus, momentum conservation is just vector addition.

10.2 Identical Particles

2. *Consider the head-on collision of two identical particles, each of mass m and energy E.*

   (a) *In Newtonian mechanics, what multiple of E is the energy E′ of one particle as observed in the reference frame of the other?*

   (b) *In special relativity, what is the energy E′ of one particle as observed in the reference frame of the other?*

   (c) *Suppose we collide two protons $(mc^2 = 1$ GeV) with energy $E = 30$ GeV. Roughly what multiple of E is E′ in this case?*

(a) In the center-of-mass frame, each particle has Newtonian kinetic energy

$$E = \frac{1}{2}mv^2. \tag{10.7}$$

In the reference frame of one of the particles, the other particle is moving twice as fast, so that

$$E' = \frac{1}{2}m(2v)^2 = 4E. \tag{10.8}$$

(b) Now we must use the relativistic energy

$$E = mc^2 \cosh \psi. \tag{10.9}$$

In the reference frame of one of the particles, the other is *not* moving twice as fast. Rather, the hyperbolic angle has doubled. Thus,

$$E' = mc^2 \cosh(2\psi) = mc^2(2\cosh^2 \psi - 1), \tag{10.10}$$

so that

$$\frac{E'}{E} = \frac{2\cosh^2 \psi - 1}{\cosh \psi}. \tag{10.11}$$

The corresponding momentum diagrams are shown in Figure 10.2 in both frames of reference.

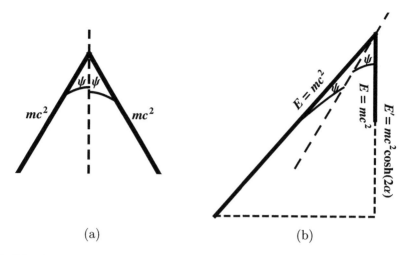

FIGURE 10.2. Momentum diagrams for the collision of identical particles, drawn both in the center-of-mass frame (left) and in the frame of
the right-hand particle (right). In both cases, the dashed line represents
the worldline of an observer in the center-of-mass frame.

(c)   We are given that

$$\cosh \psi = \frac{E}{mc^2} = 30, \qquad (10.12)$$

so that

$$\frac{E'}{E} = \frac{2 \cosh^2 \psi - 1}{\cosh \psi} \approx 60. \qquad (10.13)$$

10.3   Pion Decay I

3.   *A pion of (rest) mass m and (relativistic) momentum*
     $p = \frac{3}{4}mc$ *decays into two (massless) photons. One pho-*
     ton travels in the same direction as the original pion, and
     the other travels in the opposite direction. Find the energy
     of each photon.

First solution: We begin by determining the energy and momentum of the pion. Since $p = \frac{3}{4}mc$, we have immediately that

$$\sinh\alpha = \frac{p}{mc} = \frac{3}{4}, \tag{10.14}$$

from which it follows that

$$\cosh\alpha = \frac{5}{4} \tag{10.15}$$

by simple (hyperbolic) triangle trigonometry. Thus, the total energy is given by

$$E = mc^2 \cosh\alpha = \frac{5}{4}mc^2. \tag{10.16}$$

Since each photon satisfies $E_i = |p_i|c$, and since, without loss of generality, we can assume $p_1 > 0$ and $p_2 < 0$, conservation of energy-momentum tells us that

$$E_1 - E_2 = pc = \frac{3}{4}mc^2 \tag{10.17}$$

and

$$E_1 + E_2 = E = \frac{5}{4}mc^2, \tag{10.18}$$

from which it follows immediately that

$$E_1 = mc^2 \tag{10.19}$$

and

$$E_2 = \frac{mc^2}{4}. \tag{10.20}$$

Second solution (algebraic): Use conservation of 2-momentum $\boldsymbol{p}$, a vector quantity. From (9.33)–(9.35), we have

$$\boldsymbol{p}c = \begin{pmatrix} E \\ pc \end{pmatrix} = mc^2 \begin{pmatrix} \cosh\psi \\ \sinh\psi \end{pmatrix} = \begin{pmatrix} E_1 + E_2 \\ E_1 - E_2 \end{pmatrix}, \tag{10.21}$$

and we are given that $pc = \frac{3}{4}mc^2$, so that $\sinh\psi = \frac{3}{4}$. The computation then proceeds as in the first solution.

Second solution (geometric): Use a momentum diagram, as shown

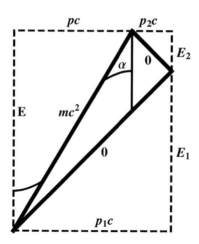

FIGURE 10.3. Momentum diagram for the decay of a pion into photons.

in Figure 10.3. The computation is the same as in the previous solutions.

Third solution: Use hyperbola geometry and triangle trigonometry, as shown in Figure 10.4 in both the given reference frame (a) and that of the pion (b), taking advantage of the fact that, in the center-of-mass frame (the rest frame of the pion), the two photons must have the same energy E_0 and equal but opposite momenta $\pm p_0 c = \pm E_0$. The details are left to the reader.

10.4   Mass and Energy

4. *A particle of mass m whose total energy is twice its rest energy collides with an identical particle at rest. If they stick together, what is the mass of the resulting composite particle, and what is its speed?*

Geometric solution: A momentum diagram for this collision is shown in Figure 10.5. The particle is shown moving to the right; its rest energy mc^2 is given by the magnitude of the hypotenuse of the right triangle shown. The observed energy, however, is $2mc^2$;

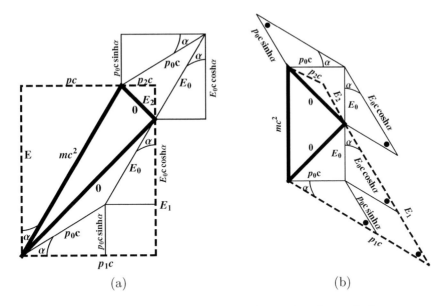

FIGURE 10.4. Analyzing pion decay by using hyperbola geometry, (a) in the laboratory frame and (b) in the rest frame of the pion.

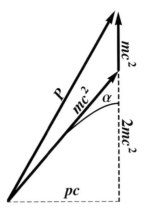

FIGURE 10.5. Spacetime diagram for the collision of identical particles.

this is the magnitude of the vertical leg of the triangle. The momentum vector of the second particle is vertical (since it is at rest), with magnitude mc^2 (the rest energy). Adding these two momentum vectors together yields the vector labeled P, whose vertical component is clearly $3mc^2$. Its horizontal component is the same as the original momentum. We have

$$\cosh \alpha = 2, \tag{10.22}$$

and therefore

$$p = mc \sinh \alpha = \sqrt{3}\, mc. \tag{10.23}$$

The rest energy of the resulting particle is therefore given by

$$M^2 c^4 = (3mc^2)^2 - (\sqrt{3}mc)^2 c^2 = 6\, m^2 c^4, \tag{10.24}$$

so that $M = \sqrt{6}m$. The speed of the resulting particle is obtained from

$$\frac{v}{c} = \tanh \psi = \frac{pc}{E} = \frac{\sqrt{3}}{3}. \tag{10.25}$$

Algebraic solution: Conservation of 2-momentum takes the form

$$mc^2 \begin{pmatrix} \cosh \psi \\ \sinh \psi \end{pmatrix} + mc^2 \begin{pmatrix} 1 \\ 0 \end{pmatrix} = \begin{pmatrix} Mc^2 \\ Pc \end{pmatrix}, \tag{10.26}$$

and we are given that $\cosh \psi = 2$. The rest of the computation proceeds as above.

10.5   Pion Decay II

5. *A pion (mass m_π) at rest decays into a muon (mass m_μ) and a massless neutrino ($m_\nu = 0$). Find the momentum p, the energy E, and the speed v/c of the muon in terms of m_π and m_μ.*

This problem can be solved by brute-force algebra; we present a more geometric solution. A momentum diagram showing this

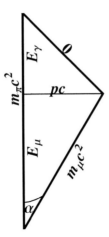

FIGURE 10.6. Spacetime diagram for the decay of a pion.

collision is given in Figure 10.6, corresponding to conservation of 2-momentum in the form

$$\begin{pmatrix} m_\pi c^2 \\ 0 \end{pmatrix} = m_\mu c^2 \begin{pmatrix} \cosh \psi \\ \sinh \psi \end{pmatrix} + \begin{pmatrix} |p_\nu|c \\ p_\nu c \end{pmatrix}. \tag{10.27}$$

The vertical line represents the pion at rest; the magnitude of this line is the pion energy $m_\pi c^2$. The muon is shown moving to the right at speed $\frac{v}{c} = \tanh \alpha$; the magnitude of the hypotenuse of the lower triangle is the muon energy $m_\mu c^2$. The upper triangle is isosceles, showing the massless neutrino; the vertical leg of this triangle is the neutrino energy $|p|c$ (since clearly $p_\nu = -p_\mu = -p$).

Conservation of energy-momentum is built into this diagram; we have, first of all, that

$$|p|c = p_\mu c = m_\mu c^2 \sinh \alpha, \tag{10.28}$$

and then that

$$m_\pi c^2 = E_\mu + |p|c = m_\mu c^2 (\cosh \alpha + \sinh \alpha) = m_\mu c^2 e^\alpha. \tag{10.29}$$

We have therefore shown that

$$e^\alpha = \frac{m_\pi}{m_\mu}, \tag{10.30}$$

so that

$$\sinh \alpha = \frac{e^\alpha - e^{-\alpha}}{2} = \frac{m_\pi^2 - m_\mu^2}{2\, m_\mu\, m_\pi}, \qquad (10.31)$$

$$\cosh \alpha = \frac{e^\alpha + e^{-\alpha}}{2} = \frac{m_\pi^2 + m_\mu^2}{2\, m_\mu\, m_\pi}, \qquad (10.32)$$

$$\tanh \alpha = \frac{\sinh \alpha}{\cosh \alpha} = \frac{m_\pi^2 - m_\mu^2}{m_\pi^2 + m_\mu^2}. \qquad (10.33)$$

Thus,

$$p_\mu = m_\mu c \, \sinh \alpha = \frac{m_\pi^2 - m_\mu^2}{2\, m_\pi} c, \qquad (10.34)$$

$$E_\mu = m_\mu c^2 \, \cosh \alpha = \frac{m_\pi^2 + m_\mu^2}{2\, m_\pi} c^2, \qquad (10.35)$$

$$\frac{v}{c} = \tanh \alpha = \frac{m_\pi^2 - m_\mu^2}{m_\pi^2 + m_\mu^2}. \qquad (10.36)$$

Alternatively, we can avoid the use of e^α by rewriting Equation (10.29) in the form

$$m_\pi - m_\mu \cosh \alpha = m_\mu \sinh \alpha. \qquad (10.37)$$

Squaring both sides of this equation and using the fundamental relationship $\cosh^2 \alpha - \sinh^2 \alpha = 1$ quickly leads to Equation (10.32), from which $\sinh \alpha$ and $\tanh \alpha$ can be obtained by using hyperbolic triangle trigonometry.

11

Relativistic Electromagnetism

In which it is shown that electricity and magnetism can no more be separated than space and time.

11.1 Magnetism from Electricity

Our starting point involves the electric and magnetic fields of an infinite straight wire, which are derived in most introductory textbooks on electrodynamics, such as Griffiths [Griffiths 99], and which we state here without proof.

The electric field of an infinite straight wire (in vacuum) with (positive) charge density λ points away from the wire with magnitude

$$E = \frac{\lambda}{2\pi\epsilon_0 r}, \qquad (11.1)$$

where r is the perpendicular distance from the wire and ϵ_0 is the permittivity constant. The magnetic field of such a wire with (positive) current I has magnitude

$$B = \frac{\mu_0 I}{2\pi r}, \qquad (11.2)$$

where r is again the perpendicular distance from the wire, and where μ_0 is the permeability constant, which is related to ϵ_0 by

$$\epsilon_0 \mu_0 = \frac{1}{c^2}. \qquad (11.3)$$

(The direction of the magnetic field is obtained as the cross product of the direction of the current and the position vector from the wire to the point in question.)

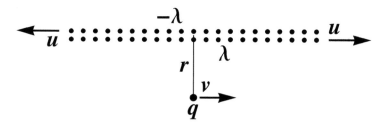

FIGURE 11.1. A wire at rest containing postive charges with charge density λ moving to the right, and negative charges with charge density $-\lambda$ moving to the left, each with speed u. A test charge q moving to the right with speed v is also shown.

We will also need the *Lorentz force law*, which says that the force $\vec{F}$ on a test particle of charge q and velocity $\vec{v}$ is given by

$$\vec{F} = q(\vec{E} + \vec{v} \times \vec{B}), \tag{11.4}$$

where $\vec{E}$ and $\vec{B}$ denote the electric and magnetic fields with magnitudes E and B, respectively.

Consider an infinite line charge consisting of identical particles of charge $\rho > 0$ separated by a distance ℓ. This gives an infinite wire with (average) charge density

$$\lambda_0 = \frac{\rho}{\ell}. \tag{11.5}$$

Suppose now that the charges are moving to the right with speed

$$u = c \tanh \psi. \tag{11.6}$$

Due to length contraction, the charge density seen by an observer at rest *increases* to

$$\lambda = \frac{\rho}{\frac{\ell}{\cosh \psi}} = \lambda_0 \cosh \psi. \tag{11.7}$$

Suppose now that, as shown in Figure 11.1, there are positively charged particles moving to the right, and equally but negatively charged particles moving to the left, each with speed u. Consider further a test particle of charge $q > 0$ situated a distance r from the wire and moving with speed

$$v = c \tanh \alpha \qquad (11.8)$$

to the right. Then the net charge density in the laboratory frame is 0, so there is no electrical force on the test particle in this frame. There is, of course, a net current density, namely

$$I = \lambda u + (-\lambda)(-u) = 2\lambda u. \qquad (11.9)$$

What does the test particle see? If we switch to the rest frame of the test particle, the negative charges appear to move faster, with speed $u_- > u$, and the positive charges appear to move slower, with speed $u_+ < u$. The relative speeds satisfy

$$\frac{u_+}{c} = \tanh(\psi - \alpha), \qquad (11.10)$$

$$\frac{u_-}{c} = \tanh(\psi + \alpha), \qquad (11.11)$$

resulting in current densities

$$\lambda_\pm = \lambda_0 \cosh(\psi \mp \alpha) = \lambda_0(\cosh \psi \cosh \alpha \mp \sinh \psi \sinh \alpha), \quad (11.12)$$

which, in turn, result in a total charge density of

$$\begin{aligned} \lambda' &= \lambda_+ - \lambda_- \\ &= -2\lambda_0 \sinh \psi \sinh \alpha \\ &= -2\lambda \tanh \psi \sinh \alpha. \end{aligned} \qquad (11.13)$$

According to Equation (11.1), this then results in an electric field of magnitude

$$E' = \frac{|\lambda'|}{2\pi \epsilon_0 r}, \qquad (11.14)$$

which in turn leads to an electric force of magnitude

$$\begin{aligned} F' = qE' &= \frac{\lambda}{\pi \epsilon_0 r} q \tanh \psi \sinh \alpha \\ &= \frac{\lambda u}{\pi \epsilon_0 c^2 r} qv \cosh \alpha \\ &= \frac{\mu_0 I}{2\pi r} qv \cosh \alpha. \end{aligned} \qquad (11.15)$$

To relate this to the force observed in the laboratory frame, we must consider how force transforms under a Lorentz transformation. We have[1]

$$\vec{F}' = \frac{d\vec{p}'}{dt'} \qquad (11.16)$$

and, of course,

$$\vec{F} = \frac{d\vec{p}}{dt}. \qquad (11.17)$$

But since in this case the force is perpendicular to the direction of motion, we have

$$d\vec{p} = d\vec{p}', \qquad (11.18)$$

and since $dx' = 0$ in the comoving frame, we also have

$$dt = dt' \cosh \alpha. \qquad (11.19)$$

Thus, in this case, the magnitudes are related by

$$F = \frac{F'}{\cosh \alpha} = \frac{\mu_0 I}{2\pi r} qv. \qquad (11.20)$$

But this is just the Lorentz force law

$$\vec{F} = q\, \vec{v} \times \vec{B} \qquad (11.21)$$

with $B = |\vec{B}|$ given by Equation (11.2).

We conclude that in the laboratory frame there is a *magnetic* force on the test particle, which is just the *electric* force observed in the comoving frame.

11.2   Lorentz Transformations

We now investigate more general transformations of electric and magnetic fields between different inertial frames. Our starting

---

[1]This is the traditional notion of force, which does not transform simply between frames. A possibly more useful notion of force is obtained by differentiating with respect to proper time, as discussed briefly in Section 11.6.

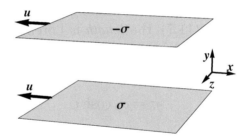

FIGURE 11.2. A capacitor moving to the left with speed u. The bottom plate has charge density $\sigma > 0$, and the top plate has charge density $-\sigma$.

point is the electromagnetic field of an infinite flat metal sheet, which is derived in most introductory textbooks on electrodynamics, such as Griffiths [Griffiths 99], and which we state here without proof.

The electric field of an infinite metal sheet with (positive) charge density σ points away from the sheet and has constant magnitude

$$E = \frac{\sigma}{2\epsilon_0}. \tag{11.22}$$

The magnetic field of such a sheet with current density $\vec{\kappa}$ has constant magnitude

$$B = \frac{\mu_0}{2}|\vec{\kappa}| \tag{11.23}$$

and direction determined by the right-hand rule.

Consider a capacitor consisting of two horizontal ($y = $ constant) parallel plates, with equal and opposite charge densities, as shown in Figure 11.2, For definiteness, take the charge density on the bottom plate to be $\sigma_0 > 0$, and suppose that the charges are at rest—that is, suppose that the current density of each plate is zero. Then the electric field is given by

$$\vec{E}_0 = E_0\,\hat{y} = \frac{\sigma_0}{\epsilon_0}\,\hat{y} \tag{11.24}$$

between the plates, and it vanishes elsewhere. Now let the capacitor move to the left with velocity

$$\vec{u} = -u\,\hat{x} = -c\tanh\psi\,\hat{x}. \tag{11.25}$$

Then the *width* of the plate is unchanged, but, just as for the line charge in Equation (11.7), the *length* is Lorentz contracted, which *decreases* the area and hence *increases* the charge density. The charge density (on the bottom plate) is therefore

$$\sigma = \sigma_0 \cosh \psi. \tag{11.26}$$

But there is now also a current density, which is given by

$$\vec{\kappa} = \sigma \, \vec{u} \tag{11.27}$$

on the lower plate. The top plate has charge density $-\sigma$, so its current density is $-\vec{\kappa}$. Then both the electric and magnetic fields vanish outside the plates, whereas inside the plates we have

$$\vec{E} = E^y \, \hat{y} = \frac{\sigma}{\epsilon_0} \, \hat{y}, \tag{11.28}$$

$$\vec{B} = B^z \, \hat{z} = -\mu_0 \, \sigma u \, \hat{z}, \tag{11.29}$$

which can be rewritten by using Equations (11.25) and (11.26), in the form

$$E^y = E_0 \cosh \psi, \tag{11.30}$$

$$B^z = B_0 \sinh \psi. \tag{11.31}$$

For later convenience, we have introduced in the last equation the quantity

$$B_0 = -c\mu_0\sigma_0 = -c\mu_0\epsilon_0 E_0 = -\frac{1}{c} E_0, \tag{11.32}$$

which does *not* correspond to the magnetic field when the plate is at rest—which, of course, vanishes since $\vec{u} = 0$.

The preceding discussion gives the electric and magnetic fields seen by an observer at rest. But what is seen by an observer moving to the right with speed $v = c \tanh \alpha$? To compute this, we first use the velocity addition law to compute the correct rapidity to insert in Equations (11.30) and (11.31), which is simply the sum of the rapidities ψ and α.

The moving observer therefore sees an electric field $\vec{E}'$ and a magnetic field $\vec{B}'$. From Equations (11.30)–(11.32), and the hyperbolic trigonometric formulas (4.5) and (4.6), we have

$$
\begin{aligned}
E'^y &= E_0 \cosh(\psi + \alpha) \\
&= E_0 \cosh\psi \cosh\alpha + E_0 \sinh\psi \sinh\alpha \\
&= E_0 \cosh\psi \cosh\alpha - cB_0 \sinh\psi \sinh\alpha \\
&= E^y \cosh\alpha - cB^z \sinh\alpha,
\end{aligned}
\tag{11.33}
$$

and similarly,

$$
\begin{aligned}
B'^z &= B_0 \sinh(\psi + \alpha) \\
&= B_0 \sinh\psi \cosh\alpha + B_0 \cosh\psi \sinh\alpha \\
&= B^z \cosh\alpha - \frac{1}{c} E^y \sinh\alpha.
\end{aligned}
\tag{11.34}
$$

The last step in each case is crucial; we eliminate ψ, E_0, and B_0, since the "lab" frame is an artifact of our construction. The resulting transformations depend, as desired, on only the two "moving" frames and their relative rapidity, α.

Repeating the argument with the y- and z-axes interchanged (and being careful about the orientation), we obtain the analogous formulas

$$
E'^z = E^z \cosh\alpha + cB^y \sinh\alpha,
\tag{11.35}
$$

$$
B'^y = B^y \cosh\alpha + \frac{1}{c} E^z \sinh\alpha.
\tag{11.36}
$$

Finally, by considering motion perpendicular to the plates we can show [Griffiths 99]

$$
E'^x = E^x,
\tag{11.37}
$$

and by considering a solenoid, we obtain [Griffiths 99]

$$
B'^x = B^x.
\tag{11.38}
$$

Equations (11.33)–(11.38) describe the behavior of the electric and magnetic fields under Lorentz transformations. These equations can be nicely rewritten in vector language by introducing the projections parallel and perpendicular to the direction of motion of the observer, namely,

$$\vec{E}_{\parallel} = \frac{\vec{v} \cdot \vec{E}}{\vec{v} \cdot \vec{v}} \, \vec{v}, \tag{11.39}$$

$$\vec{B}_{\parallel} = \frac{\vec{v} \cdot \vec{B}}{\vec{v} \cdot \vec{v}} \, \vec{v}, \tag{11.40}$$

and

$$\vec{E}_{\perp} = \vec{E} - \vec{E}_{\parallel}, \tag{11.41}$$

$$\vec{B}_{\perp} = \vec{B} - \vec{B}_{\parallel}. \tag{11.42}$$

We then have

$$\vec{E}'_{\parallel} = \vec{E}_{\parallel}, \tag{11.43}$$

$$\vec{B}'_{\parallel} = \vec{B}_{\parallel}, \tag{11.44}$$

and

$$\vec{B}'_{\perp} = \left(\vec{B}_{\perp} - \frac{1}{c^2} \vec{v} \times \vec{E}_{\perp} \right) \cosh \alpha, \tag{11.45}$$

$$\vec{E}'_{\perp} = \left(\vec{E}_{\perp} + \vec{v} \times \vec{B}_{\perp} \right) \cosh \alpha. \tag{11.46}$$

11.3 Vectors

In Chapters 5 and 9, we used 2-component vectors to describe spacetime quantities such as position and velocity, with one component for time and the other for space. In the case of three spatial dimensions, we use 4-component vectors, namely,

$$x^{\nu} = \begin{pmatrix} x^0 \\ x^1 \\ x^2 \\ x^3 \end{pmatrix} = \begin{pmatrix} ct \\ x \\ y \\ z \end{pmatrix}. \tag{11.47}$$

These are called *contravariant* vectors, and their indices are written "upstairs," that is, as superscripts.

Just as before, Lorentz transformations are hyperbolic rotations, which must now be written as 4×4 matrices. For instance, a "boost" in the x direction now takes the form

$$\begin{pmatrix} ct' \\ x' \\ y' \\ z' \end{pmatrix} = \begin{pmatrix} \cosh\alpha & -\sinh\alpha & 0 & 0 \\ -\sinh\alpha & \cosh\alpha & 0 & 0 \\ 0 & 0 & 1 & 0 \\ 0 & 0 & 0 & 1 \end{pmatrix} \begin{pmatrix} ct \\ x \\ y \\ z \end{pmatrix}. \tag{11.48}$$

A general Lorentz transformation can be written in the form

$$x'^{\mu} = \Lambda^{\mu}{}_{\nu}\, x^{\nu}, \tag{11.49}$$

where the $\Lambda^{\mu}{}_{\nu}$ are (the components of) the appropriate 4×4 matrix, and where we have adopted the *Einstein summation convention* that repeated indices, in this case ν, are to be summed from 0 to 3. In matrix notation, this can be written as

$$\boldsymbol{x'} = \boldsymbol{\Lambda}\boldsymbol{x}. \tag{11.50}$$

A new feature of Lorentz transformations in four dimensions is that some directions may be left invariant. For example, the boost above will lead to length contraction in the x direction, but not in the y or z directions. Any Lorentz transformation can be built up out of six "basis" transformations, namely the boosts in the x, y, and z directions, and the rotations about the x, y, and z axes. The Lorentz transformations form a *group*, known as the *Lorentz group*, as they are closed under multiplication and have inverses.

Why are some indices "up" and others "down"? In relativity, both special and general, it is essential to distinguish between two types of vectors. In addition to contravariant vectors, there are also *covariant* vectors, often referred to as dual vectors. The dual vector associated with x^{μ} is[2]

$$x_{\mu} = \begin{pmatrix} -x_0 & x_1 & x_2 & x_3 \end{pmatrix} = \begin{pmatrix} -ct & x & y & z \end{pmatrix}. \tag{11.51}$$

---

[2]Some authors use different conventions.

We won't have much need for covariant vectors, but note that the invariance of the interval can be nicely written as

$$x_\mu x^\mu = -c^2 t^2 + x^2 + y^2 + z^2$$
$$= x'_\mu x'^\mu. \tag{11.52}$$

(Don't forget the summation convention.) In fact, this property can be taken as the *definition* of Lorentz transformations, and it is straightforward to determine which matrices $\Lambda^\mu{}_\nu$ are allowed.

Taking the derivative with respect to proper time leads to the *4-velocity*

$$u^\mu = \frac{dx^\mu}{d\tau} = \frac{dx^\mu}{dt}\frac{dt}{d\tau}. \tag{11.53}$$

It is often useful to divide the components of the 4-velocity into space and time in the form

$$\boldsymbol{u} = \begin{pmatrix} c\gamma \\ \vec{v}\gamma \end{pmatrix} = \begin{pmatrix} c\cosh\alpha \\ \hat{v}\,c\sinh\alpha \end{pmatrix}, \tag{11.54}$$

where $\hat{v}$ is the unit vector in the direction of $\vec{v}$. Note that the 4-velocity is a unit vector in the sense that

$$\frac{1}{c^2} u_\mu u^\mu = -1. \tag{11.55}$$

The 4-momentum is simply the 4-velocity times the rest mass; that is,

$$p^\mu = mu^\mu = \begin{pmatrix} \frac{1}{c}E \\ \vec{p} \end{pmatrix} = \begin{pmatrix} mc\gamma \\ m\vec{v}\gamma \end{pmatrix} = \begin{pmatrix} mc\cosh\alpha \\ \hat{v}\,mc\sinh\alpha \end{pmatrix}. \tag{11.56}$$

Note that

$$p_\mu p^\mu = -m^2 c^2, \tag{11.57}$$

which is equivalent to our earlier result

$$E^2 - p^2 c^2 = m^2 c^4. \tag{11.58}$$

11.4   Tensors

Roughly speaking, tensors are like vectors, but with more components and hence more indices. We consider here only one particular
case, namely, *rank 2 contravariant tensors*, which have two "upstairs" indices. In a particular reference frame, the components of
such a tensor make up a 4×4 matrix,

$$
T^{\mu\nu} = \begin{pmatrix} T^{00} & T^{01} & T^{02} & T^{03} \\ T^{10} & T^{11} & T^{12} & T^{13} \\ T^{20} & T^{21} & T^{22} & T^{23} \\ T^{30} & T^{31} & T^{32} & T^{33} \end{pmatrix}. \tag{11.59}
$$

How does the tensor $\boldsymbol{T}$ transform under Lorentz transformations?
Well, it has *two* indices, *each* of which must be transformed. This
leads to a transformation of the form

$$
T'^{\mu\nu} = \Lambda^{\mu}{}_{\rho}\Lambda^{\nu}{}_{\sigma}T^{\rho\sigma} = \Lambda^{\mu}{}_{\rho}T^{\rho\sigma}\Lambda^{\nu}{}_{\sigma}, \tag{11.60}
$$

where the second form (and the summation convention) leads naturally to the matrix equation

$$
\boldsymbol{T'} = \boldsymbol{\Lambda T \Lambda}^{t}, \tag{11.61}
$$

where t denotes matrix transpose.

   Further simplification occurs in the special case where $\boldsymbol{T}$ is
antisymmetric; that is,

$$
T^{\nu\mu} = -T^{\mu\nu}, \tag{11.62}
$$

so that the components of $\boldsymbol{T}$ take the form

$$
T^{\mu\nu} = \begin{pmatrix} 0 & a & b & c \\ -a & 0 & f & -e \\ -b & -f & 0 & d \\ -c & e & -d & 0 \end{pmatrix}. \tag{11.63}
$$

11.5   The Electromagnetic Field

Why have we done all this? Well, first of all, note that, due to antisymmetry, $\boldsymbol{T}$ has precisely six independent components. Next, compute $\boldsymbol{T}'$, using matrix multiplication and the fundamental hyperbolic trig identity (4.4). As you should check for yourself, the result is

$$T'^{\mu\nu} = \begin{pmatrix} 0 & a' & b' & c' \\ -a' & 0 & f' & -e' \\ -b' & -f' & 0 & d' \\ -c' & e' & -d' & 0 \end{pmatrix} \tag{11.64}$$

where

$$a' = a, \tag{11.65}$$
$$b' = b \cosh\alpha - f \sinh\alpha, \tag{11.66}$$
$$c' = c \cosh\alpha + e \sinh\alpha, \tag{11.67}$$
$$d' = d, \tag{11.68}$$
$$e' = e \cosh\alpha + c \sinh\alpha, \tag{11.69}$$
$$f' = f \cosh\alpha - b \sinh\alpha. \tag{11.70}$$

The first three of these are similar to the transformation rule for the electric field, and the remaining three are similar to the transformation rule for the magnetic field!

We conclude that the electromagnetic field is described by an antisymmetric, rank 2 tensor of the form

$$F^{\mu\nu} = \begin{pmatrix} 0 & \frac{1}{c}E^x & \frac{1}{c}E^y & \frac{1}{c}E^z \\ -\frac{1}{c}E^x & 0 & B^z & -B^y \\ -\frac{1}{c}E^y & -B^z & 0 & B^x \\ -\frac{1}{c}E^z & B^y & -B^x & 0 \end{pmatrix} \tag{11.71}$$

which is known as the *electromagnetic field tensor*.

11.6 Maxwell's Equations

Maxwell's equations in vacuum (and in MKS units) are

$$\vec{\nabla} \cdot \vec{E} = \frac{1}{\epsilon_0}\rho, \tag{11.72}$$

$$\vec{\nabla} \cdot \vec{B} = 0, \tag{11.73}$$

$$\vec{\nabla} \times \vec{E} = -\frac{\partial \vec{B}}{\partial t}, \tag{11.74}$$

$$\vec{\nabla} \times \vec{B} = \mu_0 \vec{J} + \mu_0\epsilon_0 \frac{\partial \vec{E}}{\partial t}, \tag{11.75}$$

where ρ is the charge density, $\vec{J}$ is the current density, and the constants μ_0 and ϵ_0 satisfy Equation (11.3). Equation (11.72) is just the differential form of Gauss's law, (11.74) is Faraday's equation, and (11.75) is Ampère's law corrected for the case of a time-dependent electric field. We also have the charge conservation equation

$$\vec{\nabla} \cdot \vec{J} = -\frac{\partial \rho}{\partial t} \tag{11.76}$$

and the Lorentz force law

$$\vec{F} = q(\vec{E} + \vec{v} \times \vec{B}). \tag{11.77}$$

Maxwell's equations (11.73) and (11.74) are automatically solved by introducing the scalar potential Φ and the vector potential $\vec{A}$ and defining

$$\vec{B} = \vec{\nabla} \times \vec{A}, \tag{11.78}$$

$$\vec{E} = -\frac{\partial \vec{A}}{\partial t} - \vec{\nabla}\Phi. \tag{11.79}$$

What form do Maxwell's equations take if we rewrite them in tensor language? We consider the following derivatives of $F^{\mu\nu}$:

$$\frac{\partial F^{\mu\nu}}{\partial x^\nu} = \frac{1}{c}\frac{\partial F^{\mu 0}}{\partial t} + \frac{\partial F^{\mu 1}}{\partial x} + \frac{\partial F^{\mu 2}}{\partial y} + \frac{\partial F^{\mu 3}}{\partial z}, \tag{11.80}$$

which corresponds to four different expressions, one for each value of μ. For $\mu = 0$, we get

$$0 + \frac{1}{c}\frac{\partial E^x}{\partial x} + \frac{1}{c}\frac{\partial E^y}{\partial y} + \frac{1}{c}\frac{\partial E^z}{\partial z} = \frac{1}{c}\vec{\nabla} \cdot \vec{E} = \frac{\rho}{c\epsilon_0} = c\mu_0\rho, \quad (11.81)$$

where Gauss's law was used to get the next-to-last equality. Similarly, for $\mu = 1$, we have

$$-\frac{1}{c^2}\frac{\partial E^x}{\partial t} + 0 + \frac{\partial B^z}{\partial y} - \frac{\partial B^y}{\partial z}. \quad (11.82)$$

Combining Equation (11.82) with the expressions for $\mu = 2$ and $\mu = 3$ yields the left-hand side of

$$-\frac{1}{c^2}\frac{\partial \vec{E}}{\partial t} + \nabla \times \vec{B} = \mu_0\vec{J}, \quad (11.83)$$

where the right-hand side follows from Ampère's law. Combining these equations, and introducing the *4-current density*

$$J^\mu \doteq \begin{pmatrix} c\rho \\ \vec{J} \end{pmatrix} \quad (11.84)$$

leads to

$$\frac{\partial F^{\mu\nu}}{\partial x^\nu} = \mu_0 J^\mu, \quad (11.85)$$

which is equivalent to the two Maxwell equations with a physical source, namely, Gauss's law (11.72) and Ampère's law (11.75).

Furthermore, taking the 4-dimensional divergence of the 4-current density leads to

$$\mu_0\frac{\partial J^\mu}{\partial x^\mu} = \frac{\partial}{\partial x^\mu}\frac{\partial F^{\mu\nu}}{\partial x^\nu} = 0, \quad (11.86)$$

since there is an implicit double sum over both μ and ν, and the derivatives commute but $F^{\mu\nu}$ is antisymmetric. (We can check this by interchanging the order of summation.) Working out the components of this equation, we have

$$\frac{1}{c}\frac{\partial J^0}{\partial t} + \frac{\partial J^1}{\partial x} + \frac{\partial J^2}{\partial y} + \frac{\partial J^3}{\partial z} = 0, \quad (11.87)$$

which is just the charge conservation equation (11.76).

What about the remaining equations? Introducing the *dual tensor* $G^{\mu\nu}$ obtained from $F^{\mu\nu}$ (Equation (11.71)) by replacing $\frac{1}{c}\vec{E}$ by $\vec{B}$ and $\vec{B}$ by $-\frac{1}{c}\vec{E}$ results in

$$G^{\mu\nu} = \begin{pmatrix} 0 & B^x & B^y & B^z \\ -B^x & 0 & -\frac{1}{c}E^z & \frac{1}{c}E^y \\ -B^y & \frac{1}{c}E^z & 0 & -\frac{1}{c}E^x \\ -B^z & -\frac{1}{c}E^y & \frac{1}{c}E^x & 0 \end{pmatrix}. \tag{11.88}$$

Then the four equations

$$\frac{\partial G^{\mu\nu}}{\partial x^\nu} = 0 \tag{11.89}$$

correspond to

$$\vec{\nabla} \cdot \vec{B} = 0, \tag{11.90}$$

$$-\frac{1}{c}\frac{\partial \vec{B}}{\partial t} - \frac{1}{c}\vec{\nabla} \times \vec{E} = 0, \tag{11.91}$$

which are precisely the two remaining Maxwell equations, (11.73) and (11.74), respectively.

Thus, Maxwell's equations in tensor form are just (11.85) and (11.89)—two tensor equations rather than four vector equations.

The Lorentzian interval as written in Equation (11.52) can be thought of as the dot product "$x \cdot x$," and is a scalar invariant. Two important scalar invariants of the electromagnetic field can be constructed naturally in tensor language, namely

$$\frac{1}{2}F_{\mu\nu}F^{\mu\nu} = -\frac{1}{c^2}|\vec{E}|^2 + |\vec{B}|^2 = -\frac{1}{2}G_{\mu\nu}G^{\mu\nu} \tag{11.92}$$

and

$$\frac{1}{4}G_{\mu\nu}F^{\mu\nu} = -\frac{1}{c}\vec{E} \cdot \vec{B} \tag{11.93}$$

where we must take care with the signs of the components of the *covariant* tensors $F_{\mu\nu}$ and $G_{\mu\nu}$. These expressions can be thought of as "$\boldsymbol{F} \cdot \boldsymbol{F}$" and "$\boldsymbol{F} \cdot \boldsymbol{G}$".

Finally, it is possible to solve the sourcefree Maxwell equations by introducing a *4-potential*

$$A^\mu = \begin{pmatrix} \frac{1}{c}\,\Phi \\ \vec{A} \end{pmatrix} \tag{11.94}$$

and defining

$$F^{\mu\nu} = \frac{\partial A^\nu}{\partial x_\mu} - \frac{\partial A^\mu}{\partial x_\nu}, \tag{11.95}$$

where again we must take care with the signs of components with "downstairs" indices. Furthermore, the Lorentz force law can be rewritten in the form

$$m\frac{\partial u^\mu}{\partial \tau} = qu_\nu F^{\mu\nu}. \tag{11.96}$$

Note the appearance of the proper time τ in this equation. Just as in Chapter 9, this is because differentiation with respect to τ pulls through a Lorentz transformation, which makes this a valid tensor equation that holds in any inertial frame.

11.7   The Unification of Special Relativity

In our brief tour of relativistic electromagnetism, we have seen how special relativity unifies physical concepts previously regarded as separate. Space and time become spacetime. Energy and momentum become 4-momentum. Charge and current densities become the 4-current density. The scalar and vector potentials become the 4-potential. And, finally, the electric and magnetic fields become the electromagnetic field tensor.

12

Problems III

12.1 Vanishing Fields

1. *Suppose we know that in a particular inertial frame neither the electric field $\vec{E}$ nor the magnetic field $\vec{B}$ has an x component, but neither $\vec{E}$ nor $\vec{B}$ is zero. Consider another inertial frame moving with respect to the first one with velocity v in the x-direction, and denote the electric and magnetic fields in this frame by $\vec{E}'$ and $\vec{B}'$, respectively.*

   (a) *What are the conditions on $\vec{E}$ and $\vec{B}$, if any, and the value(s) of v, if any, such that $\vec{E}'$ vanishes for some value of v?*

   (b) *What are the conditions on $\vec{E}$ and $\vec{B}$, if any, and the value(s) of v, if any, such that $\vec{B}'$ vanishes for some value of v?*

   (c) *What are the conditions on $\vec{E}$ and $\vec{B}$, if any, and the value(s) of v, if any, such that both $\vec{E}'$ and $\vec{B}'$ vanish for the same value of v?*

   (d) *Is it possible that $\vec{E}'$ and $\vec{B}''$ vanish for different values of v?*
   (We write $\vec{B}''$ rather than $\vec{B}'$ to emphasize that $\vec{E}'$ and $\vec{B}''$ are with respect to different reference frames.)

   (a) By inserting $\vec{E}' = 0$ into Equation (11.46), we get

$$v = |\vec{v}| = \frac{|\vec{E}|}{|\vec{B}|}, \qquad (12.1)$$

   since $\vec{v}$ is perpendicular to $\vec{B}$. This is possible only if

119

$|\vec{E}| < c|\vec{B}|$, which also follows immediately from the invariance of property (11.92).

(b)   By inserting $\vec{B'} = 0$ into (11.45), we get

$$v = |\vec{v}| = \frac{c^2|\vec{B}|}{|\vec{E}|}, \qquad (12.2)$$

which is possible only if $c|\vec{B}| < |\vec{E}|$. This condition also follows immediately from the invariance of property (11.92).

(c)   This is not possible; if the electric and magnetic fields are both zero in any frame, they are zero in all frames.

(d)   No. The conditions in parts (a) and (b) cannot both be satisfied.

12.2   Parallel and Perpendicular Fields

2.   *Suppose that in a particular inertial frame, the electric field $\vec{E}$ and magnetic field $\vec{B}$ are neither perpendicular nor parallel to each other.*

(a)   *Is there another inertial frame in which the fields $\vec{E''}$ and $\vec{B''}$ are parallel to each other?*

(b)   *Is there another inertial frame in which the fields $\vec{E''}$ and $\vec{B''}$ are perpendicular to each other?*

(You may assume without loss of generality that the inertial frames are in relative motion parallel to the x-axis, and that neither $\vec{B}$ nor $\vec{E}$ has an x-component. Why? Briefly justify your assumptions.)

(a)   Yes. Setting the cross product of Equations (11.46) and (11.45) equal to zero, and using the vector identity

$\vec{u} \times (\vec{v} \times \vec{w}) = (\vec{u} \cdot \vec{w})\vec{v} - (\vec{u} \cdot \vec{v})\vec{w}$ and the fact that $\vec{v}$ is perpendicular to both $\vec{E}$ and $\vec{B}$, leads to

$$\frac{\frac{\vec{v}}{c}}{1 + \frac{|\vec{v}|^2}{c^2}} = \frac{\vec{E} \times c\vec{B}}{|\vec{E}|^2 + c^2|\vec{B}|^2}. \tag{12.3}$$

(b)   No. According to (11.93), $\vec{E} \cdot \vec{B}$ is invariant.

13

Beyond Special Relativity

NEXT STOP: General Relativity!

13.1 Problems with Special Relativity

We began in Chapter 2 by using moving trains to model inertial reference frames. But we made an implicit assumption beyond assuming an ideal train, with no friction and a perfectly straight track. We also assumed that there was no gravity. Einstein's famous thought experiment for discussing gravity is to consider a "freely falling" reference frame, typically a falling elevator. Objects thrown horizontally in such an elevator will not be seen to fall—there is no gravity (for a little while, at least). But even here there is an implicit assumption, namely, that the elevator is small compared to the earth.

Returning to our ideal train moving to the right, let's now assume there is gravity. A ball thrown straight up will, according to an observer on the train, eventually turn around and fall back down, moving along a straight line, as shown in Figure 13.1(a). Since the acceleration due to gravity at the surface of the earth is taken to be constant, the exact motion is described by a quadratic equation in t. From the ground, the same effect is seen, but combined with a constant motion (linear in t) to the right. The motion therefore takes place along a parabola; see Figure 13.2.

Now let's suppose that the train is also accelerating to the right with constant acceleration. Then a ball thrown straight up on the train still moves along the same parabolic trajectory as before (as seen from the ground), but the rear wall of the train

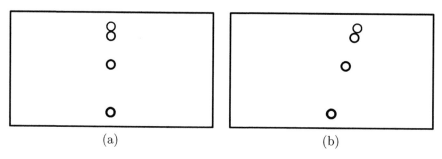

FIGURE 13.1. A ball thrown (a) straight up in a train moving at constant speed and (b) at an angle in a train moving with constant acceleration, as seen from the train.

might now catch up with it before it lands. This shows, first of all, that Newton's laws fail in a noninertial frame.

But let's analyze the situation more carefully. We may try to compensate by giving the ball's velocity a horizontal component. If the ball's initial horizontal speed is chosen appropriately, the resulting trajectory could look like the "boomerang" in Figure 13.1(b), in which (as seen from the train) the ball is thrown up at an angle and returns along the same path.

What is going on here? Acceleration and gravity produce the same kind of effect, and what the "boomerang" is telling us is that the *effective* force of gravity is no longer straight down, but rather at an angle. If we throw the ball up at that angle, it goes "up" and "down" in a straight line.

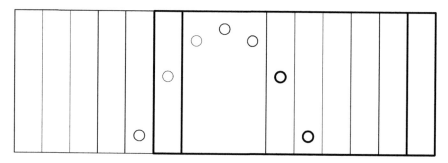

FIGURE 13.2. A ball thrown in a moving train, as seen from the ground.

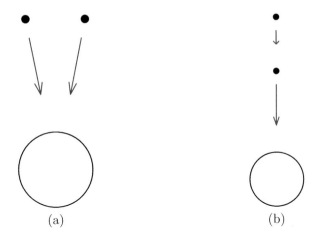

FIGURE 13.3. Two objects falling toward the Earth from far away either move (a) closer together or (b) further apart, depending on their initial configuration.

13.2 Tidal Effects

Consider now two objects falling toward the earth, but far from it, as shown in Figure 13.3(a). Both objects fall toward the center of the earth—which is not quite the same direction for each object. Assuming the objects start at the same distance from the earth, their paths will converge. Now, if they don't realize they are falling—by virtue of being in a large falling elevator, say—they will nevertheless notice that they are approaching each other. This is gravity.

Similarly, if, as shown in Figure 13.3(b), one object starts out above the other, it is slightly farther from the earth, and hence experiences a slightly weaker gravitational attraction. Thus, the lower object will always accelerate more than the upper one, and so the distance between the objects will increase.

This is the effect that causes tides. To see this, let's imagine that the earth is falling toward the moon, as shown in Figure 13.4. The earth itself can be regarded as a rigid body, falling toward the moon with the velocity given by the arrow at its center. But the

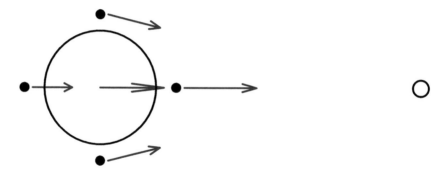

FIGURE 13.4. How tides are caused by the earth falling toward the moon.

water on the surface of the earth falls toward the moon, depending on its location, with velocities given by the four arrows shown in the figure. The relative motion of the water and the earth is given by comparing the arrows: *both* the nearest and furthest points from the moon move away from the center, causing high tides, whereas the points on either side move toward the center, causing low tides. This explains why there are two, not one, high (and low) tides every day. A similar effect, but roughly half as strong, is caused by the sun.

13.3   Differential Geometry

In two dimensions, Euclidean geometry is the geometry of a *flat* piece of paper. But there are also *curved* two-dimensional surfaces. The simplest of these is the sphere, which has constant positive curvature, and is a model of (double) *elliptic geometry*. Another important example is the hyperboloid, which also has constant curvature, and is a model of *hyperbolic geometry*. Hyperbolic and elliptic geometry form the two main categories of non-Euclidean geometries.

   In fact, *any* two-dimensional surface in Euclidean 3-space provides a possible geometry, most of which are curved. However, it is important to realize that distances are always positive in all such

signature	flat	curved
$(+ + \ldots +)$	Euclidean	Riemannian
$(- + \ldots +)$	Minkowskian	Lorentzian

TABLE 13.1. Classification of geometries.

geometries. We measure the distance between two points on such a surface by stretching a string between them *along the surface*. This does not measure the (three-dimensional) Euclidean distance between the points, however; instead, it corresponds to integrating the arc length along the shortest path between them.

In hyperbola geometry, we instead made a fundamental change to the distance function, allowing it to become negative or zero. If there is precisely one (basis) direction in which distances turn out to be negative, such geometries are said to have *Lorentzian signature*, as opposed to the *Euclidean signature* of ordinary surfaces. As implied by the way we have drawn it, hyperbola geometry turns out to be flat in a well-defined sense, which immediately raises the question of whether there are *curved* geometries with Lorentzian signature.

The mathematical study of curved surfaces forms a central part of *differential geometry*, and the further restriction to surfaces on which distances are positive is known as *Riemannian geometry*. The much more difficult case of Lorentzian signature is known, not surprisingly, as *Lorentzian geometry*, and the important special case in which the curvature vanishes is known as *Minkowskian geometry*. This classification of geometries by signature and curvature is summarized in Table 13.1.

What does this have to do with physics? We have seen that hyperbola geometry, more correctly called Minkowski space, is the geometry of special relativity. Lorentzian geometry turns out to be the geometry of general relativity. In short, according to Einstein, gravity is curvature!

13.4   General Relativity

Just as one studies flat (two-dimensional) Euclidean geometry before studying curved surfaces, one studies special relativity before general relativity. But the analogy goes much further.

The basic notion in Euclidean geometry is the distance between two points, which is given by the Pythagorean theorem. The basic notion on a curved surface is still the distance "function," but this is now a statement about infinitesimal distances. Euclidean geometry is characterized by the infinitesimal distance function

$$ds^2 = dx^2 + dy^2, \qquad (13.1)$$

which can be used to find the distance along any curve. An infinitesimal distance function such as Equation (13.1) is also called a *line element* or a *metric*.

The simplest curved surface, a sphere of radius r, can be characterized by the line element

$$ds^2 = r^2 \, d\theta^2 + r^2 \sin^2 \theta \, d\phi^2. \qquad (13.2)$$

Remarkably, it is possible to calculate the curvature of the sphere from this line element alone; it is *not* necessary to use any three-dimensional geometry. It also is possible to calculate the "straight lines"; that is, the shortest path between two given points. These are straight lines in the plane, but great circles ("equators") on the sphere.

Similarly, special relativity is characterized by the line element[1]

$$ds^2 = -dt^2 + dx^2, \qquad (13.3)$$

which is every bit as flat as a piece of paper. We get general relativity simply by considering more general line elements.

Of course, it's not quite that simple. The line element must have a minus sign. And the curvature must correspond to a physical source of gravity—that's where Einstein's field equations come

---

[1]Note that we have set $c = 1$.

in. But, given a line element, we can again calculate the "straight lines," which now correspond to freely falling observers. Matter curves the universe, and the curvature tells objects which paths are straight—that's gravity!

Here are two examples to whet the reader's appetite further. First, the (three-dimensional) line element

$$ds^2 = -dt^2 + \sin^2(t)\left(d\theta^2 + \sin^2\theta\, d\phi^2\right) \tag{13.4}$$

describes the two-dimensional surface of a spherical balloon whose radius changes with time. This roughly corresponds to a cosmological model for an expanding universe produced by a Big Bang. Studying the properties of this model more carefully leads to good questions about relativistic cosmology.

Finally, the line element

$$ds^2 = -\left(1 - \frac{2m}{r}\right)dt^2 + \frac{dr^2}{1 - \dfrac{2m}{r}} \tag{13.5}$$

describes a simplified model of a black hole, with an apparent singularity at $r = 2m$, which is, however, just due to a poor choice of coordinates. Trying to understand what actually happens at $r = 2m$ will give you some understanding of what a black hole really is.

An introductory textbook on general relativity, such as d'Inverno [d'Inverno 92] or Taylor and Wheeler [Taylor and Wheeler 00], provides further details on models such as these.

13.5 Uniform Acceleration and Black Holes

Special relativity is about inertial observers moving at constant velocity. Consider now an observer undergoing uniform acceleration. What does this mean?

We start with Newton's second law in the form

$$F = \frac{dp}{dt} = ma, \tag{13.6}$$

FIGURE 13.5. The trajectory of a uniformly accelerating object.

but we use the relativistic notion of momentum, so that

$$p = mc \sinh \alpha. \tag{13.7}$$

Setting $a = $ constant is equivalent to assuming that p is a linear function of t, leading to

$$\sinh \alpha = \frac{at}{c}. \tag{13.8}$$

But we also have

$$v = \frac{dx}{dt} = c \tanh \alpha, \tag{13.9}$$

so that

$$dx = \frac{c^2}{a} \sinh \alpha \, d\alpha, \tag{13.10}$$

and it is easy to solve this differential equation for x.

Trajectories of uniform acceleration are therefore given by

$$x = \frac{c^2}{a} \cosh \alpha \tag{13.11}$$

and

$$t = \frac{c}{a} \sinh \alpha. \tag{13.12}$$

In other words, uniform acceleration corresponds to worldlines that are hyperbolas, as shown in Figure 13.5. These hyperbolas are just our old friends—"circles" of constant distance from the origin. We now discuss the physical implications of their geometric properties.

First of all, the instantaneous speed of the accelerating object is given by the (inverse) slope of the worldline at that point. A hyperbolic worldline describes an object that is instantaneously at rest at $t = 0$, but whose speed approaches the speed of light as t goes to ∞. No, we can't actually travel at the speed of light, but with uniform acceleration we can come arbitrarily close.

The asymptotes of our hyperbolic worldline are lines at 45°, and could therefore represent beams of light. An object undergoing uniform acceleration can therefore outrun a light beam. All that is necessary is to have a head start—and to end the race after finite time.

Furthermore, the region of spacetime behind that light beam is invisible to the uniformly accelerating object. More precisely, no signal originating at $x < 0$ when $t = 0$ can ever reach the object. No light can escape this region. This is precisely the geometric idea behind black holes.

The geometry shown in Figure 13.5 is, in fact, very similar to that of the Schwarzschild black hole. "Stationary" observers far away remain at a constant distance from the black hole, despite the gravitational field; they are therefore undergoing uniform acceleration equal and opposite to that of gravity. A region of spacetime exists, "inside" the black hole, from which a light beam can never escape. Both of these properties are correctly modeled by our much simpler model involving uniform acceleration but no gravity.

14

Three-Dimensional Spacetime Diagrams

A brief introduction to special relativity in $2 + 1$ dimensions.[1]

14.1 Introduction

As we have seen, spacetime diagrams display invariant physical content geometrically. Two-dimensional spacetime diagrams are sufficient to describe most elementary applications, such as length contraction or time dilation, and even some more advanced applications, such as many of the standard paradoxes. However, some physical situations are inherently three-dimensional.

We present here several examples which benefit from the use of three-dimensional spacetime diagrams.

14.2 The Rising Manhole

The Rising Manhole problem was briefly mentioned in Section 8.4. Here is a more precise formulation:

A meter stick lies along the x-axis of the laboratory frame and approaches the origin with constant velocity. A very thin metal plate parallel to the xz-plane in the laboratory frame moves upward in the y-direction with constant velocity. The plate has a circular hole with a diameter of one meter centered on the y-axis. The center of the meter stick arrives at the laboratory origin at the same time in the laboratory frame as the rising plate arrives at the plane $y = 0$. Does the meter stick pass through the hole?

---

[1]The material in this chapter is based on the presentation in [Dray 13b].

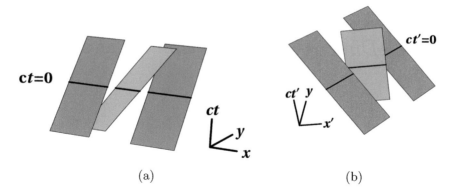

(a)   (b)

FIGURE 14.1. The worldsheets of the meter stick (center) and the metal plate (shown in two pieces with the hole in between), (a) as observed in the laboratory frame, and (b) according to the stick.

From the point of view of the metal plate (representing the manhole), the meter stick (representing the manhole cover) is very short, and should therefore easily fit through the hole. However, from the point of view of the stick, the hole in the plate is very short, so the plate should bump into the stick. Which of these scenarios is correct?

We draw a three-dimensional spacetime diagram, as shown in Figure 14.1(a) in the laboratory frame. The z-direction is suppressed; only one spatial dimension of the plate is shown. The plate is therefore represented in two pieces, with a hole in the middle; world sheets are shown for each of the two line segments in the plate along the x-axis that connect the hole to the edge of the plate. This two-part worldsheet is tilted backward, indicating motion upward (in space, that is, in the y-direction). The stick is moving toward the right, and clearly does fit through the hole, as shown by the heavy lines indicating the positions of the stick and the plate at $t = 0$. So what is observed in the stick's frame? A plane of constant time *according to the stick* is not horizontal in the laboratory frame; since this plane is tilted, so is the plate! Redraw the spacetime diagram in the reference frame of the stick, as shown in Figure 14.1(b). The worldsheet of the stick is now vertical, and the worldsheet of the plate is tilted both to the left and back. Considering the heavy lines indicating the positions of the stick and the plate at $t' = 0$ yields the traditional "snapshot" of the situation in the stick's frame; the stick is indeed longer than the hole, but passes through it at an angle.

14.3   The Moving Spotlight

Next, we return to the Moving Spotlight problem in Section 7.4.

A spaceship goes past at speed v. An antenna is mounted on its hull at an angle θ with the (horizontal) hull. A spotlight is mounted on the spaceship so that its beam also makes an angle θ with the hull. What angle θ' does an observer on the ground say the beam makes with the hull? What angle θ'' does this observer say the antenna makes with the hull?

A three-dimensional spacetime diagram in the reference frame of the spaceship is shown in Figure 14.2. The heavy line along the lightcone represents the lightray. Since the spotlight is pointed along the antenna, the lightray must lie within the worldsheet of the antenna, which is vertical, since the spaceship is at rest. The antenna is tilted toward the back of the spaceship, so that θ is measured in the direction opposite to the motion of the spaceship.

Projecting this diagram vertically yields the spatial diagram shown at the top of Figure 14.2, with the lightray superimposed on the antenna—which both clearly make the same angle with the hull. The triangle shown in this projection represents the displacements Δx and Δy traveled by the lightray in moving from one end of the antenna to the other.

Projecting the diagram horizontally, as shown at the front of Figure 14.2, yields a traditional, two-dimensional spacetime description, in which we have indicated the (projection of the) antenna at several times by horizontal lines. Since the worldsheet is not parallel to the x-axis, the lightray does *not* travel at the speed of light *in the direction the spaceship is moving*! How fast is it

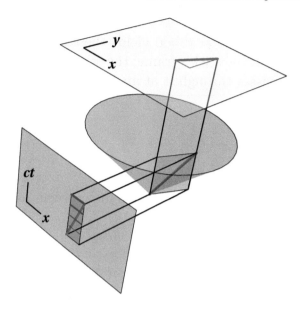

FIGURE 14.2. A lightray traveling along the tilted antenna of a moving spaceship, observed in the reference frame of the spaceship. The worldsheet of the antenna is represented by the vertical rectangle, and the heavy line along the lightcone represents the lightray. Horizontal lines represent the antenna at several different instants of time.

moving? Using the projected lightray as the hypotenuse of a right triangle, we have

$$\frac{u}{c} = \frac{\Delta x}{c\Delta t} = \tanh\psi. \tag{14.1}$$

But Δx is the same as before, and, since the (unprojected) lightray really does travel at the speed of light, we also have

$$(c\Delta t)^2 = (\Delta x)^2 + (\Delta y)^2 = \frac{(\Delta x)^2}{\cos^2\theta}, \tag{14.2}$$

so that

$$\tanh\psi = \cos\theta. \tag{14.3}$$

What is the situation as seen from the ground? As shown in Figure 14.3, the primary difference is that the worldsheet of the antenna is no longer vertical, since it is moving. But the rest of

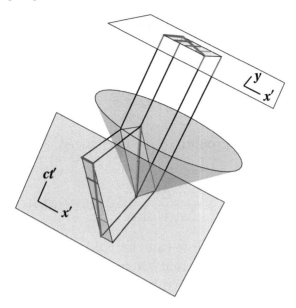

FIGURE 14.3. A lightray traveling along the antenna of a moving spaceship, according to an external observer. The worldsheet of the antenna is represented by the tilted rectangle, and the heavy line along the lightcone represents the lightray. Horizontal lines represent the antenna at several different instants of time.

the argument is the same, with $\Delta x'$, $\Delta y'$, and $\Delta t'$ again denoting the displacements of the lightray (which do not correspond to the sides of any triangles in Figure 14.3). The apparent speed of the light ray parallel to the spaceship is given by the Einstein addition law, which takes the form

$$\frac{u'}{c} = \frac{\Delta x'}{c\Delta t'} = \tanh(\alpha - \psi), \tag{14.4}$$

where $\tanh\alpha = v/c$, so that

$$\cos\theta' = \tanh(\psi - \alpha) = \frac{\tanh\psi - \tanh\alpha}{1 - \tanh\psi\tanh\alpha} = \frac{\cos\theta - \tanh\alpha}{1 - \cos\theta\tanh\alpha}, \tag{14.5}$$

or in more familiar language

$$\cos\theta' = \frac{c\cos\theta - v}{c - v\cos\theta}. \tag{14.6}$$

Determining the angle of the antenna is much easier. If the antenna is at an angle θ to the horizontal in the rest frame of the spaceship, as shown in Figure 14.4(a), then

$$\tan \theta = \frac{\Delta y}{\Delta x}. \tag{14.7}$$

As seen from the ground, y is unchanged, but x is length-contracted, so

$$\tan \theta'' = \frac{\Delta y'}{\Delta x'} = \frac{\Delta y}{\Delta x / \cosh \alpha} = \tan \theta \cosh \alpha. \tag{14.8}$$

But wait a minute. How can the lightray move along the antenna if the angles are different? Now the spatial projection at the top of Figure 14.3, reproduced in Figure 14.4(b), comes in handy, as it shows a "movie" of the lightray propagating along the (moving) antenna despite the fact that the angles between the hull and the lightray and antenna are clearly different. The same question could be raised in the Newtonian version of the problem, in which a ball is thrown along the mast of a sailboat. The resolution of this apparent paradox is the same in both contexts: Position and velocity transform differently between reference frames, and therefore so do "position angles" and "velocity angles."

14.4   The Lorentzian Inner Product

Perhaps the most famous thought experiment ever is Einstein's argument that the constancy of the speed of light leads to time dilation: A beam of light bouncing vertically in a (horizontally!) moving train travels different distances, and therefore takes different amounts of time, in the moving reference frame than in the rest frame. Thus, the "ticks" of a clock measure different time intervals in the two reference frames.

This scenario is usually drawn as a purely spatial "movie," but is inherently three-dimensional, as was briefly discussed in Section 2.5. As in Figure 6.5, Figure 14.5 shows the worldlines

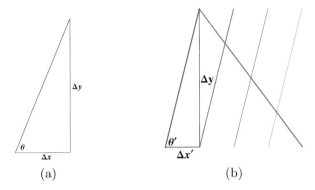

(a)   (b)

FIGURE 14.4. The horizontal projections of the spacetime diagrams in Figures 14.2 and 14.3. In the spaceship's frame (a), the antenna is at rest, but as seen from the ground (b) the spaceship moves to the left; the antenna is shown at four discrete instants of time. The beam of light moves along the antenna in both cases, as shown by the diagonal line in (b) (and not shown separately in (a)).

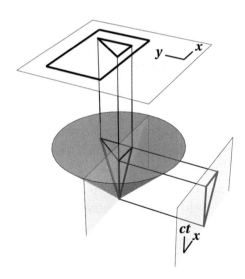

FIGURE 14.5. A bouncing lightray on a moving train. One "tick" of the lightray is represented by the diagonal line along the lightcone. The vertical projection (top) yields the usual analysis of time dilation in terms of a *Euclidean* triangle; the corresponding *hyperbolic* triangle is obtained from the horizontal projection (right).

of a lightray (along the lightcone), an observer at rest (vertical), and the lightsource on the floor of the moving train (tilted to the right).[2]

In Figure 14.5, we also show the horizontal and vertical projections. The projection into a horizontal plane $t =$ constant, shown at the top of Figure 14.5, is precisely the standard spatial "movie," shown in Figure 2.5. This triangle shows the distance $v \, \Delta t$ traveled by the train (base), the distance $c \, \Delta t$ traveled by the beam of light according to the observer on the platform (hypotenuse), and the height of the train (side), that is, the distance $c \, \Delta t'$ traveled by the beam of light according to the observer on the train. Projecting instead into the vertical plane $y = 0$, shown on the right of Figure 14.5, one obtains the spacetime diagram shown in Figure 6.6.

The simplified version of this diagram that appears in Figure 6.5 was used in Section 2.5 to derive the formula for time dilation, namely (6.9).

14.5 Transverse Directions

In three or more spacetime dimensions, there is a difference between the direction of motion, called *longitudinal*, and the remaining orthogonal spatial directions, called *transverse*.

In three dimensions, relative motion in the x-direction leaves lengths in the y direction unchanged. Thus, the two-dimensional Lorentz transformation (5.11) between two reference frames in relative motion in the x-direction becomes

$$\begin{pmatrix} x \\ y \\ ct \end{pmatrix} = \begin{pmatrix} \cosh \alpha & 0 & \sinh \alpha \\ 0 & 1 & 0 \\ \sinh \alpha & 0 & \cosh \alpha \end{pmatrix} \begin{pmatrix} x' \\ y' \\ ct' \end{pmatrix}, \qquad (14.9)$$

with an analogous transformation holding for relative motion in

---

[2]This figure also appears opposite the title page of this book, as it beautifully summarizes the geometry of special relativity!

the y-direction. As can be inferred from the manhole in Section 8.4, both the shape of an object and its spatial orientation therefore can be different in different reference frames. Squares become rectangles; horizontal becomes tilted.

But that's not all. The discussion of the spotlight at the end of Section 14.3 points out that distances and velocities transform differently in both Newtonion physics and in relativity. This difference is especially apparent when considering how velocity transforms. Just because transverse lengths are invariant does not mean that transverse velocities are!

One way to investigate this behavior is to reproduce the infinitesimal argument given in Section 9.2, which in differential form states that

$$\frac{1}{c}\frac{dx'}{dt'} = \frac{dx - \frac{v}{c}\,c\,dt}{c\,dt - \frac{v}{c}\,dx} = \frac{\frac{1}{c}\frac{dx}{dt} - \frac{v}{c}}{1 - \frac{v}{c^2}\frac{dx}{dt}}, \tag{14.10}$$

reiterating the point made there that both the numerator and denominator change. The analog of (14.10) for transverse velocity is

$$\frac{1}{c}\frac{dy'}{dt'} = \frac{dy}{c\,dt - \frac{v}{c}\,dx} = \frac{\frac{1}{c}\frac{dy}{dt}}{1 - \frac{v}{c^2}\frac{dx}{dt}}, \tag{14.11}$$

so that transverse velocities are not invariant; the numerator is unchanged, but the denominator changes.

As these examples and discussion suggest, special relativity in three dimensions introduces several new challenges compared to the two-dimensional case. As always, a combination of spacetime diagrams and careful computation can be used to resolve apparent paradoxes.

15

Minkowski Area via
Light Boxes

In which an elegant treatment of area in Minkowski space yields a
new, geometric measurement tool.[1]

15.1   Area in Special Relativity

Remarkably, the notion of area in special relativity is exactly the
same as in ordinary Euclidean geometry! Even more remarkably,
Minkowskian area can be used to measure spacetime intervals![2]

Recall that the area of a parallelogram in three-dimensional
Euclidean space is given by the magnitude of the cross product of
the vectors along two independent sides. In two dimensions, this
formula reduces to the determinant of the matrix of components
of the two vectors. Explicitly, consider the vectors $\vec{u} = u_x\hat{x} + u_y\hat{y}$
and $\vec{v} = v_x\hat{x} + v_y\hat{y}$ shown in Figure 15.1. The (signed) area A of
the Euclidean parallelogram spanned by $\vec{v}$ and $\vec{w}$ is then given by

$$A = \begin{vmatrix} v_x & v_y \\ w_x & w_y \end{vmatrix}. \tag{15.1}$$

as can easily be seen by restricting the usual formula for area in
terms of the cross product to two dimensions. In two-dimensional
special relativity, we use exactly the same definition, merely re-
placing y by t. Thus, if $\vec{v} = v_x\hat{x} + v_t\hat{t}$ and $\vec{w} = w_x\hat{x} + w_t\hat{t}$, then

---

[1] The presentation in this chapter was motivated by the discussion of causal
domains in [Salgado 16].

[2] We consider here only two-dimensional Minkowski space, where the formulas are straightforward. Our conclusions are also valid in higher dimensions,
but the computations are messier.

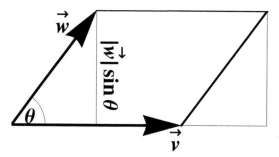

FIGURE 15.1. The parallelogram spanned by two vectors.

the area of the Minkowskian parallelogram spanned by $\vec{v}$ and $\vec{w}$ is just

$$A = \begin{vmatrix} v_x & v_t \\ w_x & w_t \end{vmatrix}. \tag{15.2}$$

It is easy to see from the definition (15.2) that, in two dimensions, Minkowskian area is nondegenerate. That is, the only parallelograms with zero area are degenerate, with all four sides parallel.[3] Furthermore, it is straightforward to show that (15.2) is invariant under Lorentz transformations; the substitutions

$$v_x \longmapsto v_x \cosh \alpha + v_t \sinh \alpha, \tag{15.3}$$
$$v_t \longmapsto v_x \sinh \alpha + v_t \cosh \alpha, \tag{15.4}$$

(and similarly for the components of $\vec{w}$) do not change the value of the determinant.

As with Euclidean area, the Minkowskian area of a rectangle is just the product of the lengths of the two sides, where by "length" of a timelike line we mean the corresponding proper time, as usual. However, all rectangles in Minkowski space have both timelike and spacelike edges. More generally, for any parallelogram, this formula does indeed reduce to the familiar "base times height," although care must be taken to ensure that these two line segments are perpendicular to each other, which is not the case for null sides, at

---

[3]This is not true in higher dimensions. However, parallelograms with at least one timelike side, or with two null sides, are still nondegenerate, which is enough for the construction below.

least in two dimensions. However, the analogous formula involving the lengths of the sides and the sine of the angle between them fails in Minkowski space, in large part because there is not always a notion of the "angle" between the sides. Furthermore, any such formula would yield zero area for any parallelogram with one or more null sides. All of these issues are neatly avoided by using the determinant (15.2) to define area.

Recall now that the squared proper time $\Delta\tau$ between two timelike-separated spacetime events is given by

$$(\Delta\tau)^2 = (\Delta t)^2 - (\Delta x)^2 = (\Delta t + \Delta x)(\Delta t - \Delta x) \qquad (15.5)$$

which is remarkable because, unlike in the Euclidean case, this expression factors. Can we find a parallelogram whose sides correspond to these factors? Yes, indeed! The area of this parallelogram will then (be proportional to) the squared spacetime interval!

We can assume without loss of generality that one of our points, P, is at the origin, and the other is given by

$$Q = (x, t) = \rho \left(\sinh \alpha \, \hat{\boldsymbol{x}} + \cosh \alpha \, \hat{\boldsymbol{t}} \right) \qquad (15.6)$$

so that the squared proper time between P and Q is just ρ^2. The secret is to construct a double-null parallelogram[4] connecting P and Q, as shown in Figure 15.2. It is straightforward to work out the components of the two null vectors, namely

$$PX = \frac{1}{2}(t - x)(-\hat{\boldsymbol{x}} + \hat{\boldsymbol{t}}) = \frac{\rho}{2} e^{-\alpha} \left(-\hat{\boldsymbol{x}} + \hat{\boldsymbol{t}} \right), \qquad (15.7)$$

$$PY = \frac{1}{2}(t + x)(\hat{\boldsymbol{x}} + \hat{\boldsymbol{t}}) = \frac{\rho}{2} e^{\alpha} \left(\hat{\boldsymbol{x}} + \hat{\boldsymbol{t}} \right). \qquad (15.8)$$

Computing the area of the parallelogram using (15.2), we obtain

$$A = \frac{1}{4} \begin{vmatrix} t + x & t + x \\ -t + x & t - x \end{vmatrix} = \frac{1}{2}(t^2 - x^2) = \frac{1}{2}\rho^2 \qquad (15.9)$$

---

[4] Wait a minute. We just agreed that the "base times height" formula doesn't apply to double-null rectangles, since the sides are not perpendicular. But the fact that they are perpendicular in Euclidean space is enough—so long as the *Euclidean* sidelengths are used—since the Euclidean and Minkowskian area formulas are formally the same!

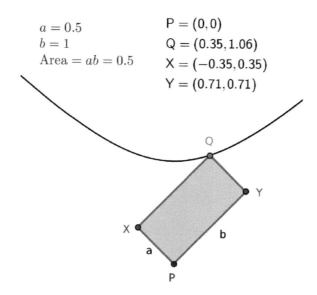

$a = 0.5$
$b = 1$
Area $= ab = 0.5$

$P = (0,0)$
$Q = (0.35, 1.06)$
$X = (-0.35, 0.35)$
$Y = (0.71, 0.71)$

FIGURE 15.2. As Q moves along the hyperbola at constant proper time ρ from P, the area of the shaded, double-null box remains constant, and is proportional to the magnitude of the squared spacetime interval between P and Q. (The Euclidean side lengths a and b are also shown; the corresponding Minkowskian side lengths are zero, but the area is nonetheless the same.)

which is proportional to the squared spacetime interval between P and Q, as claimed.

Physically, such double-null parallelograms correspond to bouncing a beam of light off of a mirror that remains at fixed distance from the worldline connecting P and Q. Such mirrors can be used as clocks along this worldline, with times expressed in multiples of the round-trip travel time for the beam of light, corresponding to one "tick" of the clock. But now we can measure time by counting ticks, each of which corresponds to a parallelogram of area $\frac{1}{2}\rho^2$, and a proper time difference of ρ. A similar argument enables the use of double-null parallelograms to measure distance.

The use of beams of light to synchonize clocks in special relativity was discussed by Einstein, and forms the heart of other geometric approaches to special relativity, including the text by

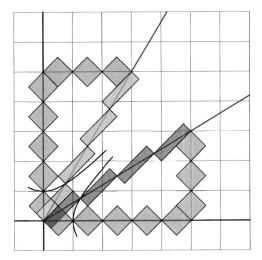

FIGURE 15.3. Using light boxes to measure time and distance.

Dragon [Dragon 12]. In a recent paper, Salgado [Salgado 16] shows how to use graph paper ruled with double-null parallelograms, which he calls "causal diamonds," to construct and analyze spacetime diagrams. The underlying justification for this approach is the area formula (15.2).

15.2 Measuring with Light Boxes

As an example of these constructions, consider the spacetime diagram shown in Figure 15.3. As usual, the vertical and horizontal axes represent the lab frame, and the tilted axes correspond to an observer moving to the right. Two 3–4–5 triangles are shown, along with the double-null parallelograms, henceforth referred to as *light boxes*.

For the vertical and horizontal sides of the triangles, the light boxes make intuitive sense; they are just fancy, two-dimensional rulers, whose diagonals can be used to measure distance and time according to the lab frame. Each such box shown has unit area (in both Euclidean and hyperbola geometry), corresponding in the timelike case to a round-trip light travel time along the edges of the box of 1 unit.

The same idea leads to the light boxes shown along the hypotenuses, corresponding to the moving observer's axes. Again, in the timelike case, the roundtrip travel time along the edges of the box is 1 unit *according to the moving observer.* This calibration can be seen explicitly using the unit hyperbolas connecting both pairs of axes.

So what have we gained? Aren't these boxes just fancy rulers?

Yes and no. As discussed in [Salgado 16], it is possible to calibrate these "clocks" experimentally, by measuring the Doppler effect. So the hyperbolas aren't actually needed. As also emphasized in [Salgado 16], the use of "rotated" graph paper, ruled with null lines (light rays) that are independent of reference frame, makes it easy to draw the light boxes for any reference frame. So yes, light boxes are fancy rulers, but their use provides a natural, unifying framework for performing frame-independent computations.

16

Hyperbolic Geometry

In which the relationship between hyperbola geometry and hyperbolic geometry is clarified.

16.1 Non-Euclidean Geometry

In two dimensions, *Euclidean* geometry is the geometry of an infinite sheet of paper. The postulates now exist in several different forms, but all address the basic properties of lines and angles. Key among them is the *parallel postulate*, which states: given any line and a point not on that line, there exists a unique line through the given point that is parallel to the given line.

It was thought by Euclid (and many after him) that the parallel postulate was so obvious that it should follow from the other postulates. In the centuries following Euclid, many incorrect proofs were proposed, purporting to show just that, but the claim is false; the parallel postulate turns out to be independent of the others. These attempted proofs ultimately gave rise in the early nineteenth century to hyperbolic geometry, the geometry in which the parallel postulate fails due to the existence of more than one parallel line.

The existence of such non-Euclidean geometries was ultimately established by considering particular models of hyperbolic geometry, several of which we discuss in this chapter. However, just as the sphere provides the most intuitive model for *elliptic* geometry, in which there are no parallel lines, the hyperboloid in Minkowski space provides the most intuitive model of hyperbolic geometry. But Minkowski space and special relativity did not exist until the early twentieth century, and were therefore not available when these other models were being developed.

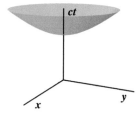

FIGURE 16.1. The hyperboloid in three-dimensional Minkowski space.

In this chapter, we run history backward: we first describe the hyperboloid model of hyperbolic geometry, and we then discuss briefly how to derive the most common traditional models of hyperbolic geometry from the hyperboloid.

16.2   The Hyperboloid

We move now to three-dimensional Minkowski space, with the squared distance function given infinitesimally (with $c = 1$) by

$$ds^2 = dx^2 + dy^2 - dt^2, \tag{16.1}$$

which can, of course, be positive, zero, or negative. We consider the hyperboloid

$$x^2 + y^2 - t^2 = -\rho^2, \tag{16.2}$$

where the radius ρ is a positive constant, and we further restrict ourselves to the branch with $t > 0$. This is the hyperboloid, as shown in Figure 16.1,

We emphasize that, just as with the sphere $x^2 + y^2 + z^2 = \rho^2$, the hyperboloid is a *two*-dimensional model, even though it lives in three dimensions because there are only two independent directions at any point *on* the hyperboloid.

To turn the hyperboloid into a geometric model, we must define its points and lines. Points are, well, points, but what should we count as "straight lines," henceforth simply called *lines*? Again, we turn to the sphere for guidance.

FIGURE 16.2. Using the hyperbolic dot product to measure distance on the hyperboloid.

The lines on a sphere are the great circles, namely, all possible equators of the sphere. We must be careful here; lines of constant longitude are great circles, but lines of constant latitude are not. Lines of constant latitude are not straight.[1] But what corresponds to an "equator" on the hyperboloid? A better characterization of lines (great circles) on the sphere is that they are the intersections of the sphere with all possible planes through the origin. This definition carries over unchanged to the hyperboloid.

How do we find the distance between two points? On the sphere, it's easy: find the angle between the corresponding vectors, and multiply by the radius. But how do we find the angle? Use the dot product. This procedure works just as well on the hyperboloid.

We now recall the formula (5.35) for the dot product of two timelike vectors. If $\vec{u}$ and $\vec{v}$ are vectors from the origin to points U and V, respectively, on our hyperboloid, then the distance between U and V is just

$$d_H(U, V) = \rho \cosh^{-1}\left(-\frac{\vec{u} \cdot \vec{v}}{\rho^2}\right). \qquad (16.3)$$

This situation is shown in Figure 16.2.

We can verify this formula in simple cases by using coordinates.

---

[1] Among other things, this fact provides a geometric explanation of why a Foucault pendulum changes its apparent plane of motion as the earth rotates.

Introducing natural "hyperboloidal" coordinates, we can write

$$U = (\rho \sinh \alpha \cos \phi, \rho \sinh \alpha \sin \phi, \rho \cosh \alpha), \qquad (16.4)$$
$$V = (\rho \sinh \psi \cos \theta, \rho \sinh \psi \sin \theta, \rho \cosh \psi), \qquad (16.5)$$

and if $\phi = 0 = \theta$, we are back in two dimensions—that is, in hyperbola geometry. In this case, the distance between U and V is just

$$\begin{aligned} d_H(U, V) &= \rho \cosh^{-1}(-\sinh \psi \sinh \alpha + \cosh \psi \cosh \alpha) \\ &= \rho \cosh^{-1}(\cosh(\alpha - \psi)) \qquad (16.6) \\ &= \rho\,|\alpha - \psi|, \end{aligned}$$

as expected.

Infinitesimal distance on the hyperboloid is obtained by rewriting expression (16.1) for infinitesimal distance in terms of hyperboloidal coordinates (ρ, α, ϕ). We obtain

$$ds^2 = \rho^2\, d\alpha^2 + \rho^2 \sinh^2 \alpha\, d\phi^2 - d\rho^2, \qquad (16.7)$$

which is not at all surprising if we recall the similar computation in spherical coordinates. Since $\rho = $ constant on the hyperboloid, infinitesimal distance on the hyperboloid is given by

$$ds^2 = \rho^2\, d\alpha^2 + \rho^2 \sinh^2 \alpha\, d\phi^2, \qquad (16.8)$$

which again recalls the line element on a sphere.

It is important to note that even though we use Minkowski space in our construction, with its indefinite distance function, the end result is a positive-definite (Riemannian) notion of distance. There are no light beams within the hyperboloid.

16.3   The Poincaré Disk

The Poincaré disk model of hyperbolic geometry is obtained from the hyperboloid by stereographic projection. Stereographic projection of an ordinary sphere maps the sphere to the xy-plane by

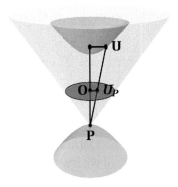

FIGURE 16.3. Stereographic projection of the hyperboloid.

projecting from the south pole; points in the northern hemisphere $(z > 0)$ map to the interior of a disk, points in the southern hemisphere $(z < 0)$ map to its exterior, and the equator $(z = 0)$ maps to the circle bounding the disk.

The idea of stereographic projection is the same for our Minkowskian hyperboloid. But what are the poles? As before, they are the two points where the t-axis intersects the hyperboloid, but now we need both branches. So the south pole is the point

$$P = (0, 0, -\rho) \tag{16.9}$$

(just as for the sphere) and stereographic projection maps each point on the $t > 0$ branch of the hyperboloid to the interior of a disk in the xy-plane.

Stereographic projection is depicted in Figure 16.3, which shows a triangle in a vertical slice through the t-axis, taken for simplicity to be in the xt-plane. The coordinates of the point U are then

$$U = (\rho \sinh \alpha, 0, \rho \cosh \alpha), \tag{16.10}$$

and the use of similar triangles tells us that the image point U_P has coordinates

$$U_P = \left(\frac{\rho \sinh \alpha}{1 + \cosh \alpha}, 0, 0 \right). \tag{16.11}$$

The image of the general point U, with coordinates as given by (16.4), would be

$$U_P = (R \cos \phi, R \sin \phi, 0), \qquad (16.12)$$

where

$$R = \frac{\rho \sinh \alpha}{1 + \cosh \alpha} \qquad (16.13)$$

is the radial coordinate in the Poincaré disk. It is now a straight-forward exercise in algebra to verify that the line element on the hyperboloid, rewritten in terms of R, is just

$$ds^2 = \frac{4 \rho^4 \left(dR^2 + R^2 \, d\phi^2\right)}{(\rho^2 - R^2)^2}, \qquad (16.14)$$

in agreement with the literature for the line element of the Poincaré disk (which normally assumes $\rho = 1$).

We have therefore shown by direct computation that the Poincareé disk model can be thought of as a relabeling of the points on the hyperboloid; these models are completely equivalent. The Poincaré disk has the twin advantages of living in two dimensions and not requiring Minkowski space for its construction, but the hyperboloid has the advantage of sharing many obvious symmetries with the sphere.

So what is the Poincaré disk *model*? As before, a geometric model is specified by giving its points and lines. Points are points, namely the interior of the unit disk in the xy-plane. But what are the lines? Since the lines on the hyperboloid are contained in planes through the origin, but stereographic projection uses lines form the south pole, the answer to this question is not immediately obvious. It turns out that *lines* in the Poincaré disk model are arcs of *Euclidean* circles that intersect the boundary of the disk orthogonally (at both points of intersection). A special case is, of course, radial lines, corresponding to circles of infinite radius. Some hyperbolic lines are shown in Figure 16.4,

A closely related model is the *Poincaré half-plane*, which is obtained from the Poincaré disk by using a complex fractional linear transformation, namely,

$$X + iY = -2 \frac{i(x + iy) + \rho}{(x + iy) + \rho i}, \qquad (16.15)$$

FIGURE 16.4. Some hyperbolic lines in the Poincaré disk.

where $x + iy = Re^{i\phi}$. Another exercise in algebra verifies that this transformation maps the disk of radius ρ onto the half-plane $Y \geq 0$, and further that

$$ds^2 = \frac{4\rho^4 (dx^2 + dy^2)}{(\rho^2 - x^2 - y^2)^2} = \rho^2 \frac{dX^2 + dY^2}{Y^2} \tag{16.16}$$

(where $x^2 + y^2 = R^2$), which again agrees with the literature (for $\rho = 1$).

The points in the Poincaré half-plane model of hyperbolic geometry are the points in the xy-plane with $y > 0$. The lines are open, Euclidean semicircles of any radius centered at points along the x-axis, including vertical lines, which correspond to an infinite Euclidean radius.

Both the Poincaré disk and Poincaré half-plane models are *conformal* models in that their line elements are multiples of the ordinary Euclidean metric. Since all distances are merely rescaled, angles can be defined through the dot product and are unaffected by the rescaling, which cancels out. Thus, angles in both of these models can be measured with Euclidean protractors.

16.4   The Klein Disk

The Klein disk model of hyperbolic geometry is also a projection of the hyperboloid. This time, however, we project from the origin

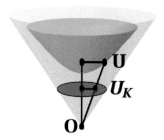

FIGURE 16.5. Constructing the Klein disk.

into the plane $t = \sigma$. Thus, the point U, with coordinates as given by (16.4), would be mapped to the point

$$U_K = (\sigma \tanh \alpha \cos \phi, \sigma \tanh \alpha \sin \phi, \sigma), \qquad (16.17)$$

as shown in Figure 16.5. The image of the hyperboloid is again the interior of a disk—this time, the disk of radius σ in the plane $t = \sigma$. The constant σ can lie between 0 and ρ, as shown in the figure, although it is useful to choose $\sigma = \rho$, in which case the disk is actually tangent to the hyperboloid at its tip.

Since lines on the hyperboloid are contained in planes through the origin, and since planes intersect in straight (Euclidean) lines, the Klein disk model has the nice feature that lines are Euclidean; the lines of the Klein disk model are the chords of the circle bounding the disk, as shown in Figure 16.6. This feature is at first sight much nicer than any of the previous models, in which "lines" bear little resemblance to Euclidean lines. However, this simplicity comes at a price. More algebra reveals that the hyperboloid line element, rewritten by using coordinates from the Klein disk, takes the form

$$ds^2 = \frac{\rho^4 \, dK^2}{(\rho^2 - K^2)^2} + \frac{\rho^2 \, K^2 \, d\phi^2}{(\rho^2 - K^2)}, \qquad (16.18)$$

where

$$K = \rho \tanh \alpha \qquad (16.19)$$

FIGURE 16.6. Some hyperbolic lines in the Klein disk.

is the radial coordinate in the Klein disk. Due to the different denominators in the two terms, the line element is no longer conformal to the Euclidean metric. Therefore, angles *cannot* be measured with a Euclidean protractor in the Klein disk model.

16.5   The Pseudosphere

Each of the preceding models of hyperbolic geometry discussed in this chapter has its advantages and disadvantages. The two Poincaré models are conformal, so angles are easy to measure, but lines are not obvious. Lines in the Klein model are obvious, but angles are difficult to measure. And for all its similarity with the sphere, in the hyperboloid model neither lines nor angles are easy to determine. Furthermore, none of these models exists in ordinary three-dimensional Euclidean space. Perhaps that is no surprise; after all, why should a model for hyperbolic geometry be constructible within Euclidean geometry? The surprise is that this is indeed possible, as we now show.

The pseudosphere is most simply related to the Poincaré half-plane, which we described in terms of dimensionless variables (X,Y) in Section 16.3. We now introduce new variables ψ and

θ via

$$\sin \psi = \frac{1}{Y}, \tag{16.20}$$

$$\theta = X, \tag{16.21}$$

and compute

$$ds^2 = \rho^2 \frac{dX^2 + dY^2}{Y^2} = \rho^2 \left(\cot^2 \psi \, d\psi^2 + \sin^2 \psi \, d\theta^2 \right). \tag{16.22}$$

We claim that this line element describes a surface of revolution in Euclidean space.

A surface of revolution in ordinary three-dimensional Euclidean space can be described in cylindrical coordinates (r,θ,z) by specifying r and z in terms of some parameter ψ. The line element on the surface is then

$$ds^2 = dr^2 + r^2 \, d\theta^2 + dz^2 = \left(\left(\frac{dr}{d\psi} \right)^2 + \left(\frac{dz}{d\psi} \right)^2 \right) d\psi^2 + r^2 \, d\theta^2.$$

$$\tag{16.23}$$

Comparing Equations (16.23) and (16.22), we must have

$$r = \rho \sin \psi, \tag{16.24}$$

and therefore,

$$\left(\frac{dz}{d\psi} \right)^2 = \rho^2 \cot^2 \psi - \rho^2 \cos^2 \psi = \rho^2 \frac{\cos^4 \psi}{\sin^2 \psi}. \tag{16.25}$$

One solution of this differential equation is

$$z = -\rho \cos \psi - \rho \ln \tan \left(\frac{\psi}{2} \right). \tag{16.26}$$

Equations (16.24) and (16.26) with $\psi \in (0, \pi/2]$ describe a curve known as a *tractrix*, as shown in Figure 16.7(b). The *pseudosphere* is the surface of revolution obtained by rotating the tractrix about the z-axis, as shown in Figure 16.7(a). We can also embed the pseudosphere in one of the other models of hyperbolic geometry, as shown for the hyperboloid in Figure 16.8. The periodicity in

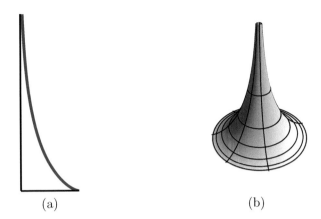

(a) (b)

FIGURE 16.7. (a) The tractrix, whose surface of revolution is the pseudosphere (b).

θ requires that we identify the edges as shown in the figure; we can think of the pseudosphere as a "rolled-up" piece of the hyperboloid. We can then think of the horizontal slices of the pseudosphere as circles centered at infinity (the tip of the pseudosphere), with infinite radius (the length from the center, as measured along the pseudosphere).

But what is so special about the pseudosphere?

A Euclidean plane has the property that it is the same everywhere, and in all directions. Thus, we can place a piece of cardboard on a flat table anywhere, and in any orientation, and it will fit snugly. We describe this by saying that the table is flat (as is the piece of cardboard); its curvature is zero. A sphere has the same property. For example, if we peel an orange, we can put a piece of the peel back anywhere on the orange, in any orientation, and it will fit snugly. The orange has constant (positive) curvature. The hyperboloid also has this property; it has constant negative curvature. If you could peel it, the peel would fit snugly anywhere, and in any orientation. But the hyperboloid lives in Minkowski space, so this property seems to be far removed from reality.

Yet we have succeeded in building a model of the hyperboloid in Euclidean space. This model has exactly the same intrinsic

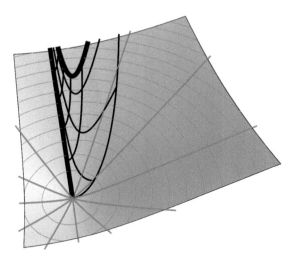

FIGURE 16.8. Embedding the pseudosphere in the hyperboloid. The heavy lines are to be identified.

properties as any other model of hyperbolic geometry, including the hyperboloid. If we were to build a pseudosphere, and then "peel" it,[2] we could take that piece of peel and put it anywhere, in any orientation, and it would fit snugly, without tearing or stretching.

This is indeed remarkable.

_____

[2] One way to do this would be to use the pseudosphere as a mold to make a partial copy, for instance with a spray-on rubber-like product. Robert Osserman, a well-known geometer, has done this demonstration during public lectures.

17

Calculus

In which a simple derivation of the derivatives of trigonometric and exponential functions is presented, without the use of limits, numerical estimates, solutions of differential equations, or integration.[1]

17.1 Circle Trigonometry

We consider once again the point P on the circle of radius r with coordinates $(r\cos\theta, r\sin\theta)$, as shown in Figure 3.2, and we recall the geometric definitions of the basic trigonometric functions (3.2) and (3.3).

We would like to compute the derivatives of these functions. What do we know? We know that (infinitesimal) arc length along the circle is given by

$$ds = r\,d\theta, \tag{17.1}$$

but we also have the (infinitesimal) Pythagorean theorem, which tells us that

$$ds^2 = dx^2 + dy^2. \tag{17.2}$$

Furthermore, from $x^2 + y^2 = r^2$, we obtain

$$x\,dx + y\,dy = 0. \tag{17.3}$$

Putting this information together, we have

$$r^2\,d\theta^2 = dx^2 + dy^2 = dx^2\left(1 + \frac{x^2}{y^2}\right) = r^2\frac{dx^2}{y^2}, \tag{17.4}$$

---

[1]The material in this chapter is based on the presentation in [Dray 13a].

so that

$$d\theta^2 = \frac{dx^2}{y^2} = \frac{dy^2}{x^2}, \tag{17.5}$$

where we have used Equation (17.3) in the last step. By carefully using Figure 3.2 to check signs, we can take the square root of Equation (17.5) and rearrange terms to obtain

$$dy = x\,d\theta,$$
$$dx = -y\,d\theta. \tag{17.6}$$

Finally, inserting (3.2) and (3.3) and using the fact that $r = $ constant, we recover the familiar expressions

$$d\sin\theta = \cos\theta\,d\theta,$$
$$d\cos\theta = -\sin\theta\,d\theta. \tag{17.7}$$

We have thus determined the derivatives of the basic trigonometric functions from little more than the geometric definition of those functions and the Pythagorean theorem—and the ability to differentiate simple polynomials.

17.2 Hyperbolic Trigonometry

The derivation in Section 17.1 carries over virtually unchanged to hyperbolic trigonometric functions. We recall from Section 4.3 the geometric definitions of the hyperbolic trigonometric functions (4.13) and (4.14), as shown in Figure 4.2.

We compute the derivatives of these functions by using the same technique as before. We choose to work on the hyperbola $y^2 - x^2 = \rho^2$, which is the hyperbola shown in Figure 4.2 that contains the point B. What do we know? We know that the (infinitesimal) arc length along the hyperbola is given by

$$ds = \rho\,d\alpha, \tag{17.8}$$

but we also have the (infinitesimal, Lorentzian) Pythagorean theorem, which tells us that

$$ds^2 = dx^2 - dy^2. \tag{17.9}$$

Furthermore, from $y^2 - x^2 = \rho^2$, we obtain

$$x\,dx = y\,dy. \tag{17.10}$$

Putting this information together, we have

$$\rho^2\,d\alpha^2 = dx^2 - dy^2 = dx^2\left(1 - \frac{x^2}{y^2}\right) = \rho^2\,\frac{dx^2}{y^2}, \tag{17.11}$$

so that

$$d\alpha^2 = \frac{dx^2}{y^2} = \frac{dy^2}{x^2}, \tag{17.12}$$

where we have used Equation (17.10) in the last step. By carefully using Figure 4.2 to check signs, we can take the square root of Equation (17.12) and rearrange terms to obtain

$$dy = x\,d\alpha,$$
$$dx = y\,d\alpha. \tag{17.13}$$

Finally, by inserting definitions (4.13) and (4.14) and using the fact that $\rho = \text{constant}$, we obtain

$$d\sinh\alpha = \cosh\alpha\,d\alpha,$$
$$d\cosh\alpha = \sinh\alpha\,d\alpha. \tag{17.14}$$

We have thus determined the derivatives of the basic hyperbolic trigonometric functions.

17.3 Exponentials (and Logarithms)

Hyperbolic trigonometric functions are usually defined by using formulas (4.1) and (4.2), and it takes some work (and independent knowledge of the exponential function) to show directly that our definition is equivalent to this one. We turn this on its head and instead *define*

$$\exp(\alpha) = \cosh\alpha + \sinh\alpha. \tag{17.15}$$

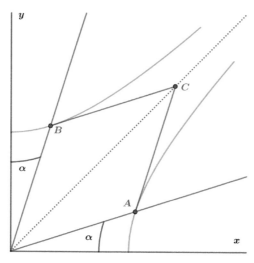

FIGURE 17.1. The geometric definition of $\exp(\alpha)$. Points A and B are as in Figure 4.2, so that the coordinates of point C are $(\rho \exp(\alpha), \rho \exp(\alpha))$.

The function $\exp(\alpha)$ as defined in (17.15) has an immediate geometric interpretation, as shown in Figure 17.1, where it is important to recall that α is the *hyperbolic* angle, and not the Euclidean angle that would be measured by a (Euclidean) protractor.

We immediately have

$$d\Big(\exp(\alpha) \Big) = \sinh \alpha \, d\alpha + \cosh \alpha \, d\alpha = \exp(\alpha) \, d\alpha, \qquad (17.16)$$

but it remains to show that $\exp(\alpha)$ as defined above is really the same as e^{α}. It is enough, however, to show that $\exp(\alpha)$ is *an* exponential function, that is, that it satisfies

$$\exp(0) = 1, \qquad (17.17)$$
$$\exp(u + v) = \exp(u) \exp(v), \qquad (17.18)$$

and to then *define*

$$e = \exp(1). \qquad (17.19)$$

We can then verify that this definition of e agrees with the usual one in several ways, such as by resorting to uniqueness results for the solutions of differential equations.

First, however, we need to verify conditions (17.17) and (17.18). The first of these, (17.17), follows immediately from the fact that $\alpha = 0$ corresponds to putting point A at $(\rho, 0)$ in Figure 4.2. Examining condition (17.18), we see that we need to know the addition formulas for the hyperbolic trigonometric functions, which we now construct.[2]

We first construct right triangles whose legs are not horizontal and vertical. We recall that the derivative of a vector of constant magnitude must be perpendicular to that vector, since

$$\vec{v} \cdot \vec{v} = \text{constant} \implies 0 = d(\vec{v} \cdot \vec{v}) = 2\,\vec{v} \cdot d\vec{v}. \qquad (17.20)$$

Since the position vector to points on a circle centered at the origin has constant magnitude, we have shown that radii are orthogonal to circles. This argument holds just as well in hyperbola geometry, and establishes orthogonality between the radial lines intersecting our hyperbolas, and the tangent lines to the hyperbolas at the point of intersection.

We consider now the situation shown in Figure 17.2. In Figure 17.2(a), angle ψ is in standard position, leading to the right triangle shown, with radial leg $\rho \cosh \psi$ and horizontal leg $\rho \sinh \psi$. In Figure 17.2(b), angle ψ has been shifted to the right, but since the hypotenuse and radial leg of the resulting right triangle have the same length as before, so must the remaining leg; the triangles in the two drawings are congruent in hyperbola geometry.

We now redraw Figure 17.2(b) as shown in Figure 17.3, so that point B corresponds to angle α, and point A corresponds to angle $\psi + \alpha$ (as measured from the y-axis). Thus, the coordinates of these points are given by

$$B = (\rho \sinh \alpha, \rho \cosh \alpha),$$
$$A = \Big(\rho \sinh(\psi + \alpha), \rho \cosh(\psi + \alpha)\Big). \qquad (17.21)$$

We then introduce point C as shown in Figure 17.3, so that the

---

[2]As in the Euclidean case, an argument that uses the geometric and algebraic properties of the Lorentzian dot product is possible, but it is less intuitive.

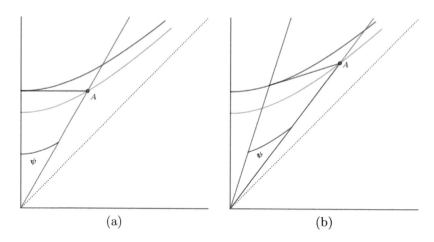

FIGURE 17.2. Using a shifted right triangle to represent the hyperbolic trigonometric functions. (a) Angle ψ is in standard position. (b) Angle ψ is shifted to the right.

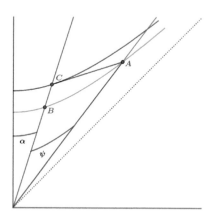

FIGURE 17.3. The geometric construction of the hyperbolic addition formulas.

distance from the origin to C is $\rho \cosh \alpha$; that is, so that the co-ordinates of C are given by

$$C = B \cosh \psi. \tag{17.22}$$

But by the discussion of Figure 17.2, the other leg of the triangle has length $\rho \sinh \psi$, so we must also have

$$C = A - B^{\perp} \sinh \psi, \tag{17.23}$$

where

$$B^{\perp} = (\rho \cosh \alpha, \rho \sinh \alpha) = \frac{dB}{d\alpha} \tag{17.24}$$

denotes the direction tangent to the hyperbola at B. By combining Equations (17.22) and (17.23), we find that

$$A = B \cosh \psi + B^{\perp} \sinh \psi, \tag{17.25}$$

and we have established the addition formulas[3]

$$\sinh(\psi + \alpha) = \sinh \psi \cosh \alpha + \cosh \psi \sinh \alpha,$$
$$\cosh(\psi + \alpha) = \cosh \psi \cosh \alpha + \sinh \psi \sinh \alpha. \tag{17.26}$$

Returning to Equation (17.18), we finally have

$$\begin{aligned}
\exp(u + v) &= \cosh(u + v) + \sinh(u + v) \\
&= \cosh(u) \cosh(v) + \sinh(u) \sinh(v) \\
&\quad + \sinh(u) \cosh(v) + \cosh(u) \sinh(v) \\
&= \Big(\cosh(u) + \sinh(u) \Big) \Big(\cosh(v) + \sinh(v) \Big) \\
&= \exp(u) \exp(v), \tag{17.27}
\end{aligned}$$

as desired.

A direct geometric verification is also possible, based on the construction in Figure 17.1, and as shown in more detail in Figure 17.4. Denoting the origin by O, and noting that line AD is

---

[3]The reader is encouraged to work out the analogous argument for ordinary trigonometry, especially if the argument about "shifted" triangles seems unclear.

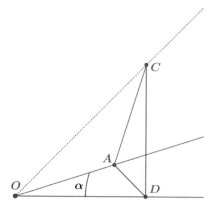

FIGURE 17.4. The geometric verification that "exp" is exponential.

at $45°$, we see that the ratio of the length of OD to that of OA is precisely $\exp(\alpha)$. In this sense, "rotation" through α (taking A to D) corresponds to stretching by a factor of $\exp(\alpha)$. Composing two such "rotations" leads directly to condition (17.18); the details are left to the reader.

Having verified the desired properties of the exponential function, it is now straightforward to define logarithms as the inverse of exponentiation, that is, to define $\log(u)$ by

$$v = \log(u) \iff u = \exp(v) \tag{17.28}$$

and to establish that this definition leads to the usual properties of the natural logarithm.

Bibliography

[d'Inverno 92] Ray A. d'Inverno. *Introducing Einstein's Relativity.* Oxford: Oxford University Press, 1992.

[Dragon 12] Norbert Dragon. *The Geometry of Special Relativity – a Concise Course.* Berlin, Heidelberg: Springer-Verlag, 2012.

[Dray 90] Tevian Dray. "The Twin Paradox Revisited." Am. J. Phys. **58** (1990), 822–825.

[Dray 03] Tevian Dray. "The Geometry of Special Relativity." Course notes, Oregon State University, 2003.

[Dray 04] Tevian Dray. "The Geometry of Special Relativity." Physics Teacher (India) **46** (2004), 144–150.

[Dray 10] Tevian Dray. "The Geometry of Special Relativity." Available at `http://physics.oregonstate.edu/coursewikis/GSR`, 2010.

[Dray 13a] Tevian Dray. "Using Differentials to Differentiate Trigonometric and Exponential Functions." College Math. J. **44**, 17–23 (2013). Available at `http://www.math.oregonstate.edu/bridge/papers/trig.pdf`.

[Dray 13b] Tevian Dray. "Using Three-dimensional Spacetime Diagrams in Special Relativity." Am. J. Phys. **81**, 593–596 (2013). Available at `http://physics.oregonstate.edu/coursewikis/GSR/book/updates/3dajp`.

[Dray 17] Tevian Dray. "The Geometry of Relativity." Am. J. Phys. **85**, 683–691 (2017).

[Dray and Manogue 06] Tevian Dray and Corinne A. Manogue. "The Geometry of the Dot and Cross Products." JOMA **6** (June 2006). Available at `http://www.joma.org/?pa=content&sa=viewDocument&nodeId=1156`.

[Einstein 1905] Albert Einstein. "Zur Elektrodynamik bewegter Körper." *Annalen der Physik*, **322** (1905), 891–921. Available in German at `http://wikilivres.info/wiki/Zur_Elektrodynamik_bewegter_K%F6rper`. Available in English at `http://en.wikisource.org/wiki/On_the_Electrodynamics_of_Moving_Bodies`.

[Griffiths 99] David J. Griffiths. *Introduction to Electrodynamics*, Third edition. Upper Saddle River, NJ: Prentice Hall, 1999.

[Norton 06] John D. Norton. "Discovering the Relativity of Simultaneity: How Did Einstein Take 'The Step'." In: *Einstein in a Trans-cultural Perspective*, edited by Yang Jiang and Liu Bing. Beijing: Tsinghua University Press, to appear. Available at `http://www.pitt.edu/~jdnorton/papers/AE_1905_Tsinghua.pdf`, or online with animated graphics at `http://www.pitt.edu/~jdnorton/Goodies/rel_of_sim`.

[Salgado 16] Roberto B. Salgado. "Relativity on Rotated Graph Paper." Am. J. Phys. **84** (2016), 344–359.

[Taylor and Wheeler 63] Edwin F. Taylor and John Archibald Wheeler. *Spacetime Physics*. San Francisco: W H Freeman, 1963.

[Taylor and Wheeler 92] Edwin F. Taylor and John Archibald Wheeler. *Spacetime Physics*, Second edition. New York: W H Freeman, 1992.

[Taylor and Wheeler 00] Edwin F. Taylor and John Archibald Wheeler. *Exploring Black Holes: Introduction to General Relativity*, San Francisco: Addison-Wesley, 2000.

Index